Springer Theses

Recognizing Outstanding Ph.D. Research

For further volumes:
http://www.springer.com/series/8790

Aims and Scope

The series "Springer Theses" brings together a selection of the very best Ph.D. theses from around the world and across the physical sciences. Nominated and endorsed by two recognized specialists, each published volume has been selected for its scientific excellence and the high impact of its contents for the pertinent field of research. For greater accessibility to non-specialists, the published versions include an extended introduction, as well as a foreword by the student's supervisor explaining the special relevance of the work for the field. As a whole, the series will provide a valuable resource both for newcomers to the research fields described, and for other scientists seeking detailed background information on special questions. Finally, it provides an accredited documentation of the valuable contributions made by today's younger generation of scientists.

Theses are accepted into the series by invited nomination only and must fulfill all of the following criteria

- They must be written in good English.
- The topic should fall within the confines of Chemistry, Physics, Earth Sciences, and related interdisciplinary fields such as Materials, Nanoscience, Chemical Engineering, Complex Systems and Biophysics.
- The work reported in the thesis must represent a significant scientific advance.
- If the thesis includes previously published material, permission to reproduce this must be gained from the respective copyright holder.
- They must have been examined and passed during the 12 months prior to nomination.
- Each thesis should include a foreword by the supervisor outlining the significance of its content.
- The theses should have a clearly defined structure including an introduction accessible to scientists not expert in that particular field.

Kirsten von Elverfeldt

System Theory in Geomorphology

Challenges, Epistemological Consequences and Practical Implications

Doctoral Thesis accepted by
The University of Vienna, Austria

Author
Dr. Kirsten von Elverfeldt
Institut für Geographie und Regionalforschung
Fakultät für Wirtschaftswissenschaften
Alpen-Adria-Universität Klagenfurt
Universitätsstrasse 65-67
9020 Klagenfurt
Austria

Supervisor
Prof. Dr. Thomas Glade
Institut für Geographie und Regionalforschung
Geomorphologische Systeme und Risikoforschung
Universität Wien
Universitätsstr. 7
1010 Wien
Austria

This book is a translation of the thesis in German "Systemtheorie in der Geomorphologie—Problemfelder, erkenntnistheoretische Konsequenzen und praktische Implikationen"

ISSN 2190-5053
e-ISSN 2190-5061
ISBN 978-94-007-2821-9
e-ISBN 978-94-007-2822-6
DOI 10.1007/978-94-007-2822-6
Springer Dordrecht Heidelberg New York London

Library of Congress Control Number: 2011946219

Printed on acid-free paper

Springer is part of Springer Science+Business Media (www.springer.com)

The journey is the reward.

Parts of chapters 1, 3, 4 and 5 are based on or taken from the publications listed below

Elverfeldt, Kirsten von & Keiler, Margreth (2008): Offene Systeme und ihre Umwelt—Systemperspektiven in der Geomorphologie. In: Heike Egner, Beate M.W. Ratter und Richard Dikau (Eds.), Umwelt als System—System als Umwelt? Systemtheorien auf dem Prüfstand. Oekom, München, pp. 75–102.

Egner, Heike & Elverfeldt, Kirsten von (2009): A bridge over troubled waters? Systems theory and dialogue in geography. *Area*, 41(3): 319–328.

Elverfeldt, Kirsten von & Glade, Thomas (2011): Systems theory in Geomorphology. A challenge. *Zeitschrift für Geomorphologie*, 55(3): 87–108.

Supervisor's Foreword

Investigations on system theory within geomorphology are still rare. However, there is a long history of working on "system theory" and "geomorphology" (e.g. Chorley 1962). Unfortunately, in the German speaking countries such as Austria, Germany and Switzerland, profound and systematic analysis on this topic are missing. This is in clear contrast to personal statements or contributions within discussion rounds where one often hears, that a "system-theory" thinking is very evident within geomorphology. Consequently there is seen no need for such studies.

This presented research shows the contrast. Indeed, the "thinking" in terms of "system theory" is often inherent in traditional geomorphic analysis. However, it is often not made explicit. And this makes the difference. One of the major underlying problems is that there is neither a common scientific ground for "system theory in geomorphology" nor a sound and founded reflection of the relevant theoretical concepts. As a consequence, there are no common definitions on relevant terms available.

Therefore, Kirsten von Elverfeldt digs in a very muddy, sticky and heavy ground. Despite the many difficulties starting with the general problem that there is no recent textbook available on system theory in geomorphology, Kirsten managed to start right from the scratch. Indeed, there are some publications available (e.g. Chorley (1962), Scheidegger (1992), Thorn and Welford (1992), Phillips (2011) and Dikau (2006)), however, it appeared to be necessary to explore the topic on "system-theory" scientifically. Embedded in international publications, Kirsten explored the current situation. The research investigates the different understandings and mirrors these by reviewing other important disciplines also working scientifically on "system-theory", such as sociology, physics, biology, and socio-ecology, to name the most important ones only. Based on this theoretical framework, various applications are investigated in detail. Practical implications and recommendations finalize this research study.

This research can be regarded as a significant contribution in the field of "geomorphic system theory", which indeed deserves widespread attention. Besides the advances in conceptual, technical and modelling fields of geomorphology, this research is—in my opinion—definitely on the leading edge in the field of "system theory". I am really looking forward to the response of the scientific community, internationally but also and in particular in the German speaking countries. I wish us all a new and innovative impulse to continue our scientific discussion in geomorphology, not only in a purely scientific theoretical debate, but also in real applications. The work of Kirsten von Elverfeldt might substantially contribute to it.

University of Vienna Prof. Dr. Thomas Glade

Preface

To write a preface is a difficult task. Prefaces are always a balancing act, as they offer insight into the author's personality. For this dissertation thesis, I have read many books, and from time to time the biggest pleasure was to read the prefaces and/or epilogues. Often, I even burst out laughing (or, quite the opposite, put the book aside). Subsequently, the author somehow resonated in the back of my mind, and thus reading the book was a better and, yes, more personal venture.

What was my motivation to write a dissertation, which deals in such depth and width with geomorphological system theory? Probably, the first academic roots for this are in Richard Dikau's working group in Bonn, where there was no way around dealing with theory. This tradition has also been continued by Thomas Glade in Vienna. However, I obviously neglected or forgot to look at the bigger picture—it was pretty comfortable in my world of geomorphology. Suddenly, however, I was pulled out of this comfortable little world by a seminar taught by Heike Egner: I got to know Niklas Luhmann's system approach. And—for whatever reason—I allowed myself to feel irritated and started to ask myself (and others!) uncomfortable questions. And as I subsequently discovered, at some stage I had obviously crossed a *point of no return*: I simply was not able to think as I had done before. This was the starting point of my dissertation and I began to delve into theories far beyond geomorphology. Despite the fact that, in the beginning, I had no notion at all of where this would lead me, I soon figured out that these theories always brought me back to geomorphology (despite some fierce self-doubts during that process). But even more so, and this is the most enriching and fascinating aspect, with each of these theories I have always learned something 'for life'. It was a pleasure.

For this thesis to be a success—as I hope—, I first and foremost owe thanks to my two mentors Thomas Glade and Heike Egner. In some sense, they have created a force field of holding and driving forces in the centre of which (or should I say: equilibrium of which?) this thesis came into existence. I owe thanks to you, Thomas, for giving me the freedom to pursue this thesis; I know that this has not always been easy. And I owe thanks to you, Heike, for bringing the joy of science back to me.

Of course, I also want to sincerely thank several other people, first of all the members of the working group ENGAGE at the department of geography and regional science of Vienna University. But particularly, I want to thank those friends, family members, and colleagues who took time for proofreading: Rainer Bell, Heike Egner, Brigitta von Elverfeldt, Christine Embleton-Hamann, Melanie Kappes, Margreth Keiler, Ronald Pöppl, Peter Weichhart and Eva Zelzer. My dear friend Oliver Löhmer always provided plenty of impulses with our physics discussions, and he also proved that physics can be a very humorous business. Walter Lang helped considerably with the figures. Karen Meehan was an enormous help for the translation of my thesis—thank you! My family and friends have supported, distracted, encouraged, and cheered me up in many ways. I am so happy that you are in my life! My deepest thanks, however, go to my son, who always reminded me that the essential things in life are somewhere else.

Contents

Chapter 1
(System-)Theoretical Thinking: A Challenge to Geomorphology?

> *I do think that there exists at least one philosophical problem which is of interest to all thinking humans. It is the problem of cosmology: The problem to understand the world—including ourselves and our knowledge as part of this world.*
>
> Karl Popper 1959

We encounter the term "system" literally everywhere, in everyday discussions about the newest communication systems just as in scientific discussions about the "system earth", the climate system, the social system, or questions regarding the political system (cf. [1, 2]). But where does system thinking come from, and even more so: Does "the" system thinking exist at all?

The contemporary system theoretical vocabulary such as "complexity", "thresholds", "self-organised-criticality" and "feedbacks" is of a relatively young age and is rooted in developments which have taken place since the 1940s and 1950s. Still, already Aristotle's widely cited sentence "the whole is more than the sum of its parts" is a prequel of an explicit system's thinking ([2, p. 19]). In general, the ancient Greeks considered a system to be an orderly set whole, whereas from the seventeenth century onwards within theology and philosophy system was rather seen as a set of theories. Since then, i.e. the baroque era, the word system quite explosively became a vogue word, thereby losing terminological strictness and unambiguousness [3]. This certain arbitrariness survived the following centuries, despite several attempts to counteract this development.

As diverse as the notion of system was understood through time, as continuous, on the other hand, is a discordance kept: The question, whether system thinking principally is organismic or mechanistic [3]. Systems, e.g. a state as a whole, the human body or the economy, from time to time have rather been understood as machine (e.g. as a watch) or as an organism. It was Immanuel Kant who in 1790 stressed an important difference:

> In a watch, one part is the tool for the movement of the other parts, but a wheel is never the acting cause for the creation of another; any part indeed exists for the other, but not through the other ([4, p. 280], translation by KvE).

This means that a watch (or a part of it) will never create another watch (or another part of it),

K. von Elverfeldt, *System Theory in Geomorphology*,
Springer Theses, DOI: 10.1007/978-94-007-2822-6_1,

> or even corrects itself if it got into disorder: all of which, however, we can expect from organised nature. Nature, "in fact, organises itself in a way self-preservation requires accordingly to the respective circumstances" ([4, p. 280f]).

In consequence, this does also mean that natural systems, according to Kant, are intentional, i.e. working towards self-preservation. Despite Kant's argument, the old antagonism between "mechanism" as synonym for a reductionist approach and "organism" as synonym for a holistic approach as foundation for explaining the world have persisted to the present. According to this understanding, geomorphological system theory is mechanistic, i.e. reductionist.

Hence, today's system theories have to be viewed in a broad historical context. Their younger roots are founded in developments which took place in the 1950s and which have to be largely attributed to the Austrian biologists Ludwig von Bertalanffy (1901–1972) and Paul Alfred Weiss (1898–1989) (cf. [5]). According to Bertalanffy's general systems theory, organisms should be regarded as open systems, which, for their metabolism, absorb mass and energy from their environment (cf. e.g. [6, 7]). Furthermore, Bertalanffy also suggests intentionality, as open systems as he understands them are striving for a steady state (Fließgleichgewicht). Although the general system theory was intended to be a description and explanation of living beings, Bertalanffy thought that his theory and its basic assumptions were universally applicable to all systems (cf. [8, 9]). Bertalanffy's approach was a revolution with the aim to unify the separated sciences. The centre of this revolution was interdisciplinarity ([10, p. 109]).

The claim to be generally applicable to all system might be, on the other hand, also the reason why the system theory has been split up in probably just as many concepts as there are disciplines. These concepts cannot be consolidated into one single theory, which stands against Ludwig von Bertalanffy's real intention. After all it has to be stated that Bertalanffy did not succeed in finding a formulation of universal principles which are applicable to all systems (cf. [10]).

Ludwig von Bertalanffy also considered his general systems theory to be an attempt to overcome the old antagonism between "mechanism" and "organism". Despite all his efforts to include holistic approaches, Bertalanffy's general systems theory is rooted in a mechanistic and causality-focussed thinking [3]. This mechanistic keynote also predominates newer (biological) system theories which, however, partly is intended as deliberate provocation and aversion of animistic/vitalistic tendencies[1] which often accompany the organismic view on systems (cf. [11–13]). In this case the equation of (living) systems with machines is quite deliberate and intends to focus on a network of processes which bring forth system elements. Mechanistic in this context means that only forces and principles are used which can be found in the physical world (cf. [14, pp. 180–183]).

System theoretical concepts which rely on those of Ludwig von Bertalanffy were introduced to geomorphology more than fifty years ago [15–19], even though the

[1] Animism or vitalism is characterized by the assumption that the multitude of living systems has to be attributed to a creative power, or that nature bears in itself a vital force.

notion of "system" has been used in geomorphology before (e.g. [20, p. 124]). The breakthrough of the then innovative concept, however, only came in the 1970s with the textbook by Richard Chorley and Barbara Kennedy (21), who adopted Bertalanffy's ideas and rethought them for geomorphology. The potential of the general systems theory for geomorphology lay in the fact that the focus was shifted away from single details and the rather descriptive explanation of their evolution towards a systematic analysis of contexts within the physical world. Furthermore, system theory offered a suitable theoretical framework for the increasing number of quantitative studies within geomorphology: In- and outputs of geomorphic systems can be quantified, as well as the throughput of mass (and energy) within the systems. Hence, the import and adaption of the general systems thinking and the accompanying change of the "geomorphological view" have been the sine qua non for last century's fundamental developments in geomorphology such as sediment budgeting and process geomorphology, but also global environmental change research–especially, if change is to be quantified. Furthermore, physical geographers hoped that systems theory would turn out to be a "unifying methodology" ([22, p. 538]) for geography which, however, has not proven to be true yet. Quite the contrary: Whenever physical and humangeographers discuss the relationship between mankind, society, and nature it quickly becomes clear that these discussions cannot be conducted based on the respective system theories, as they are as different as day and night. There are differences in what is accepted to be regarded as system and to which extent system theories are a suitable resource for geographical explanations. Albeit (or rather: because) the terminology is partly identical, a mutual understanding is difficult to attain (an attempt in this direction can be found in [23–25], as the meaning of the terms are partly contrary. Hence, there seems to be a general agreement (at least on the level of scientific-political declarations) that some of today's essential scientific and social challenges—e.g. global change and natural hazards and risks—are more successfully dealt with in an innergeographical cooperation. However, both part-disciplines have developed contrasting theoretical frameworks, hampering serious cooperation. Among other reasons, this is due to the fact that systems theory did not succeed in becoming an equally important background theory in human geography.

Within human geography there was a short phase in the 1970s in which there was an attempt to apply systems theory to human-geographic problems (c.f. e.g. [26–28]). One of the main reasons, however, why general systems theory and systems thinking in general was not successfully established in social sciences was the mechanistic conception of the world as mentioned above. Still, despite this early failure, human geography also assumes that there are certain entities in society (e.g. economy, science, or organisations) which can be addressed as systems. In contrast to geomorphology, however, there have never been any efforts to establish a methodology for system analysis. However, approximately since the beginning of the twentieth century there has been an increasing development within (German) human geography to anchor a new system theoretical approach (cf. e.g. [29–32]). This approach was developed by the German sociologist Niklas Luhmann (1927–1998) and is based on recent system-theoretical concepts such as autopoiesis, self-organisation, and self-reference which have been developed and discussed in the natural sciences, mainly biology and physics, since the 1970s and 1980s. Niklas Luhmann united these ideas, transferred them to the social sciences, and thus attained his theory of social systems [33–39].

These new concepts—it has to be stressed again—had been developed within the natural sciences such as the system-theoretical concepts of the biologists Humberto Maturana and Francisco Varela (1946–2001) [40] and form the basis on which Luhmann built his theory of social systems. The differences between these concepts and the traditional system theory are far-reaching. An essential difference is that these theories are no longer based on the assumption of open systems which are striving for a steady state or a dynamic equilibrium with their environment as it was postulated by Bertalanffy (e.g. [9, 41]). Furthermore, these new theories are formulated within the paradigm of self-organisation (cf. e.g. [42–45]) and they emphasize the "becoming" instead of system states [46, 47]. Therefore, they require a different perception—and a complex, network-like school of thought rather than thinking in straight lines and causalities (cf. [48]). Additionally, the term "system" is no longer tied to a specific structure, or to a changing configuration of specific components, or even to a specific constellation of internal or external relations. On the contrary: The term "system" now stands for the coherence of evolving, interactive clusters of processes which manifest themselves from time to time in globally stable structures, but which have nothing to do with any equilibrium or the solidity of technical structures ([43, p. 32]).

Thus, two system-theoretical approaches can be distinguished: On the one hand there exist system-theoretical approaches based on the general systems theory of Ludwig von Bertalanffy which focus on in- and outputs, steady states, and linear relations between single components. These approaches are labelled as first order system theories, and one example for such a first order system theory is the geomorphological system theory. Second order system theories, on the other hand, can be distinguished from first order system theories by extending the classical concepts by including aspects like autopoiesis, self-reference, and self-organisation. The approaches in biology [40, 49–53], and in sociology [34, 36, 54], but also in cybernetics [55–57], and physics [43, 47, 58–65] can be classified as such second order system theories.

From this new view of the world as expressed in these second order system theories it follows that other aspects than before are perceived: For example, an increasing number of self-organizing processes have been found, and terms like autopoieses, dissipative structures and fractals have not only been coined, but also transferred to a variety of scientific disciplines ([45, p. 63]). Science now has to face this paradigm shift, and this is precisely the challenge geomorphology also has to meet.

In order to answer the question about how big this challenge might be and to what extent a rethinking is indispensable the geomorphological research practice and its theoretical basis has to be observed and viewed. This is ventured within this book. For this I utilized a method suggested by Heinz von Foerster's observation theory [66], which will be detailed in Chap. 2. It is, in this case, the observation that a scientific discipline ascribes itself a specific power of observation, i.e. the "geographic view" or, even more specific, the "geomorphic view" as it is spread in lectures and conference talks. However, the insight that observations do not speak for themselves is not very common in geography and geomorphology. Quite often, in geomorphology theory is seen as nonessential or as subordinate to field experience:

"More important [...] is the fact that geomorphology is and probably always will be a field-oriented science [...][and] geomorphic insight [...] must be acquired gradually through long field experience"([67, p. 3]).

It is this low perception of theoretical basics which is repeatedly criticized by some geomorphologists who ask their colleagues to see the bigger picture behind their observations and to recognize their own theoretical presumptions, as otherwise the observations remain nothing but an non-interpretable pattern of impressions [68]. Rhoads and Thorn ([69, p. 51]) also emphatically stress that all observations are theory dependent. The same is made clear by Hard ([70, p. 41]) who emphasizes that even the observers who deem themselves as being free of any theory make a theoretical-metatheoretical decision which they consider as that self-evident and correct that they do not even know that they made such a decision. To conclude, one basic problem of geomorphology is its "epistemological imperative of empiricism" (Dikau 2008, personal communication), or, as Chorley [71] has set it:

> Whenever anyone mentions theory to a geomorphologist, he instinctively reaches for his soil auger.

Why, however, does it seem to be that difficult within physical geography to accept (or at least to thoroughly discuss) the idea of the theory-dependency of all observations or to ascribe an equal importance to theory as to empirics? Possibly this can be explained by the history of physical geography, as it was already an applied science in the eighteenth/nineteenth century, at a time of the great explorers such as Humboldt. Physical geography then served as resource description and exploration and as regional research. In this sense, physical geography has never been a "pure" science which knowledge had to be brought to application (cf. [72, p. 118]). Furthermore, this could explain why geographers are commonly trained not to (and especially not in advance) engage in theory and theoretical discussions, but rather to "just go there" in order to see (unbiasedly and naïve just as through the eyes of a child) "what's actually up" in the field ([73, p. 315]).

Closely related to this attitude is the observation that physical geographers seem to be predestined to blunder into the "ontologication-trap" [3], i.e. mistaking the wish that the research object functions as a system with a real characteristic of the system. In doing so, the thing we are looking at becomes a system "in itself". Hard ([70, p. 47]) even goes one step further in his description of geographical praxis and locates a typical figure of thought which he calls the "twofold ontological non sequitur": In his opinion, geographers generally not only project theory upon reality and reify it there (first "ontological non sequitur"), but they then also easily infer in the reverse from this reality structure the correct method (etc.) (second "ontological non sequitur"). If you're interested in testing this statement by Hard, just start a discussion with students, phds, post-docs, and professors in lectures and at conferences—Hard's analysis is still very up to date, although the exceptions might have increased within the last ten years.

Fig. 1.1 A landscape (Langental, Dolomites, Southern Tyrol), in which geomorphologists can read? The 'right' or 'wrong' can only be seen retrospectively and even then, it stays contingent (Photo: T. Glade)

As a matter of course, within geomorphology it is assumed that we—provided we have received a 'good' training–can downright read the landscape (also cf. Fig. 1.1). From this point of view, the 'correct' reading is absolutely essential, as only then e.g. hotspots in the countryside can be detected, i.e. spots which will (always) pose a hazard to society (like the San Andreas Fault in California) (also cf. [74, 75]). Goudie ([76], p 247), for example, refers to this ability as "having an eye for the country". The essential question in this context is, however: When do we know that we have 'read' correctly? This can only be judged retrospectively, and even then this judgement is only temporary and contingent, as it depends on the point of view (presuppositions, background theory, knowledge, research focus, ...) of the observer: The judgement will *always* be made against the background of a specific theory. In this sense, therefore, this book is geomorphic to the core, as it on the one hand offers the methodical requisite know-how for recurring retrospectives (observation theory, Chap. 2), and on the other hand it offers a theoretical framework in which previous research can be adjudicated upon.

Geomorphological practicing thus has been critically self-observed in the past as well as in the present, and this self-observation[2] has revealed a specific basic

[2] In my opinion this self-observation is a huge credit of the discipline. However and strangely, this self-observation does not lead to changes of the common research practice within geomorphology. Pointedly one could say: We know about our weaknesses but do not do anything about them.

attitude of "the typical" geomorphologist: That is, geomorphologists learn most by being in the field and not by reading books. And, as a logical consequence, system theory not only always has been, but still is a challenge to geomorphology. This book meets the challenge and is a contribution to a deliberate theoretical reflection within geomorphology. Thus, the main underlying hypothesis of this book is: Theoretical performance offers geomorphology a considerable surplus value, as

(1) stringent definitions of the research object are only possible with a sound and strong theoretical foundation. Without such definitions we simply do not know what we are doing. Questions arising from this are: How stringent and structuring are current basic definitions in geomorphology? And if there are logical errors, where are they? Are there any alternatives?
(2) without theory we have no means to reduce the world's complexity. However, only by this reduction, science becomes possible. The degree to which we perceive the world as complex, confusing, or boundless reveals quite a lot about our basic scientific organizing structure. How is the world perceived within geomorphology and how big is the organizing potential of our background theory?
(3) theoretical performance ensures geomorphology as a (natural) science stays in touch which physics. Hence, theory allows for a solid foundation for the practice and, at the same time, for a new aplomb in dealing with our neighbouring disciplines. Hence, the question is: How firmly is geomorphological theory and practice rooted in physics? And if there are any inconsistencies, what are the reasons for these and how can they be counteracted?
(4) theoretical performance requires that we think about how we detect geomorphological issues in general. It is the question already mentioned above: Can we read in the landscape, and do the issues simply force themselves on us, or do we perceive the world accordingly to our specific internal structures (our cognitive apparatus, our experiences, our school of thought etc.)? It is an epistemological question. If we critically discuss the geomorphological background theory, epistemology (and with it the question of the range of our scientific results) cannot be excluded.
(5) even in a discipline as strongly empirically oriented as geomorphology the question is not whether to conduct empirics or theory. The question is: How strong is the theoretical basis on which empirics is grounded? And if the basis changes, what does this mean for empirics?

The focus of this book thereby lies upon geomorphological systems theory as it is assumed that this is the relevant background theory. The aim of this book therefore is to expatiate the theoretical basis of geomorphology and to stringently and coherently make available developments of other systems theories. In this sense I do agree with Gerhard Hard ([70], p 45), who pointed out that we do not have the choice whether or not we practice meta-theory. We only have the choice to do it (at least partially) in a controlled and conscious manner or to do it in an underhand manner and blindly.

The questions mentioned above span diverse problem areas which will be discussed in Part 1 of the book (Chaps. 3–8). The first problem area (Chap. 3) is caused by the geomorphic tradition to mostly handle theory implicitly. Hence, the coherence and stringency of geormorphological basic assumptions and terminology is weak. Furthermore, geomorphological system theory implies an understanding of systems by which literally everything can be regarded as system. Second-order system theories, especially those from biology, offer an alternative to this traditional understanding as system boundaries are given by a system-specific process group which is involved in the self-creation (autopoiesis) of the respective system. This process group can only be found within the system and not within its environment, thus providing a simple and unequivocal criterion for the definition of systems.

Within this chapter, the discussion is limited to the second-order system theories of Maturana & Varela as well as Luhman in order to avoid confusion and because they offer the most explicit definitions of what can be addressed as system.

A second problem area (Chap. 4) of geomorphological system theory arises from the basic assumption that geomorphic systems are principally open. This assumption is based on a specific world view according to which everything is connected to everything else. I will show that as a consequence of this view systems seem to be fully determined by the boundary (environmental) conditions, and that the system behaviour seems to be causally linked to these environmental conditions. On the contrary, second-order system theories, especially those of biology, offer concepts like self-organisation, autopoiesis, self-reference, and operative closeness which may deliver explanations as to why systems do not always act according to the (changed) environmental conditions. Furthermore, this phenomenon of self-organised behaviour of systems builds a central aspect of non-equilibrium thermodynamics, i.e. the physical basis of geomorphological system-theoretical analyses.[3]

The physical basics are the subject of Chap. 5: Those physical theories in which geomorphology is (or should) be rooted are presented. As thermodynamics is the basis for any macroscopic analyses of physical systems, thus including geomorphic systems, this will be the foundation of all discussions. The focal point will be placed on non-equilibrium thermodynamics, and mainly on the theory of dissipative structures by Ilya Prigogine. The comparison of the basic assumptions in physics and geomorphology will reveal that the physical basis of geomorphology resembles shaky grounds rather than a solid fundament: Too often, differing, incompatible scopes from mechanics and thermodynamics are mixed. Therefore, geomorphological statements often exceed their limited scope of application.

In Chap 6 the geomorphological equilibrium concept is soundly analysed and discussed. The incoherencies and logical errors of this concept will firstly be analysed for themselves, and secondly as compared to the physical basis.

---

[3] Thermodynamics deals with the question of which forms of energy exist and in how far they are capable of doing work. Per definition, thermodynamics is a systems theory as it is macroscopic in scope and therefore always focuses on the relationships of different components and the processes and the energy exchange between them. Hence, it is the current paradigm under which natural (non-living) systems, including geomorphic systems are observed.

Another problem area—that of nonlinearity and complexity (Chap. 7)—partly results from the problems mentioned above, but also from the geomorphological habitus to assign theory minor importance. Therefore, developments from other disciplines are indeed taken on, whilst there is nearly no discussion on own theoretical fundamentals and schools of thought. Consequently, geomorphology shows a coexistence and hotchpotch of several incompatible basic assumptions and diversely defined terms, with the result that not only can geomorphology's research object (i.e. the landscape) be described as palimpsest, but also the discipline itself (cf. [77]). This is the reason why complexity theories which—with their focus on discontinuities and nonlinearity—should function as antithesis to equilibrium thinking (cf. [78]) are utilized within the equilibrium framework in geomorphology. Building upon the previous chapters the theoretical and terminological roots of this contradiction will be discussed.

The final chapter (Chap. 8) of Part I summarizes similarities and differences of the various system theories. Furthermore, the advantages which will result for geomorphology from an adaptation of some of the concepts are presented.

Part II focuses on the epistemological and practical implications of an adaptation of second-order system theories to geomorphology (Chaps. 9 and 10). The conclusion (Chap. 11) shows to what extent it is possible to modify geomorphological system theory according to the concepts from physics, biology, and even sociology in order to meet the challenges and problems which have been discussed in Part I.

In the next chapter, to start this venture, the method which was used to analyse the theoretical basis of geomorphology will be presented.

References

1. Egner H, von Elverfeldt K (2009) A bridge over troubled waters? Systems theory and dialogue in geography. Area 41(3):319–328
2. Müller K (1996) Allgemeine Systemtheorie. Geschichte, Methodologie und Sozialwissenschaftliche Heuristik eines Wissenschaftsprogramms. Westdeutscher Verlag, Opladen, p 381
3. Müller R (2009): Geschichte des Systemdenkens und des Systembegriffs (http://www.muellerscience.com/SPEZIALITAETEN/System/systemgesch.htm, 15 Dec 2009)
4. Immanuel K (2001) Kritik der Urteilskraft. Felix-Meiner Verlag, Hamburg, p 535
5. Drack M, Apfalter W (2006) Is Paul Weiss' and Ludwig von Bertalanffy's system thinking still valid today?, The 50th Annual Meeting of the International Society for the Systems Sciences, pp 1–10
6. von Bertalanffy L (1940) Der Organismus als physikalisches System betrachtet. Die Naturwissenschaften 28(33):521–531
7. von Bertalanffy L (1954) Das Fließgleichgewicht des Organismus. Kolloid-Zeitschrift 139:86–91
8. von Bertalanffy L (1950) The theory of open systems in physics and biology. Science 111(2872):23–29
9. von Bertalanffy L (1950) An outline of general systems theory. The British J Philosophy Sci 1(2):134–165
10. Dubrovsky V (2004) Toward system principles: general system theory and the alternative approach. Systems Research and Behavioral Science 21:109–122

11. Bühl WL (1987) Grenzen der Autopoiesis. Kölner Zeitschrift für Soziologie und Sozialpsychologie 39:225–254
12. Maturana HR (1982) Die Organisation des Lebendigen: eine Theorie der lebendigen Organisation. In: Maturana HR (ed) Erkennen: Die Organisation und Verkörperung von Wirklichkeit. Ausgewählte Arbeiten zur biologischen Epistemologie. Wissenschaftstheorie, Wissenschaft und Philosophie. Vieweg, Braunschweig, pp 138–156
13. Maturana HR, Varela JF (1982) Autopoietische Systeme: Eine Bestimmung der lebendigen Organisation. In: Maturana HR (ed) Erkennen: Die Organisation und Verkörperung von Wirklichkeit. Ausgewählte Arbeiten zur biologischen Epistemologie. Wissenschaftstheorie, Wissenschaft und Philosophie. Vieweg, Braunschweig, pp 170–235
14. Maturana HR, Varela JF, Uribe G (1982) Autopoiese: die Organisation lebender Systeme, ihre nähere Bestimmung und ein Modell. In: Maturana HR (ed) Erkennen: Die Organisation und Verkörperung von Wirklichkeit. Ausgewählte Arbeiten zur biologischen Epistemologie. Wissenschaftstheorie, Wissenschaft und Philosophie. Vieweg, Braunschweig, pp 157–169
15. Chorley JR (1962) Geomorphology and general systems theory. Geological Survey Professional Paper. United States Government Printing Office, Washington, pp 1–10
16. Hack J T (1960): Interpretation of erosional topography in humid temperate regions. Am J Sci, Bradley Volume, 258-A: 80–97
17. Strahler AN (1950) Equilibrium theory of erosional slopes approached by frequency distribution analysis. Part I Am J Sci 248:673–696
18. Strahler AN (1950b) Equilibrium theory of erosional slopes approached by frequency distribution analysis. Part II: Significance tests applied to slope problems in the Verdugo and San Rafael Hills, California. American Journal of Science, 248: 800–814
19. Strahler AN (1952) Dynamic basis of geomorphology. Bulletin of the Geological Society of America 63:923–938
20. Gilbert GK (1877) Geology of the Henry Mountains. Government Printing Office, Washington, p 160
21. Chorley RJ, Kennedy BA (1971): Physical geography–a systems approach, London
22. Stoddart DR (1967) Organism and ecosystem as geographical models. In: Richard JC, Haggett P (eds) Models in geography. Methuen, London, pp 511–548
23. Egner H (2008a) Gesellschaft, Mensch, Umwelt–beobachtet. Ein Beitrag zur Theorie der Geographie. Erdkundliches Wissen. Franz Steiner, Stuttgart, p 208
24. Egner H (2008) Komplexität. Zwischen Emergenz und Reduktion. In: Egner H, Ratter BMW, Dikau Richard (eds) Umwelt als System–System als Umwelt? Systemtheorien auf dem Prüfstand. oekom, München, pp 39–54
25. Egner H (2008) Planen, beeinflussen, verändern. Zur Steuerbarkeit autopoietischer Systeme. In: Egner H, Ratter BMW, Dikau R (eds) Umwelt als System–System als Umwelt? Systemtheorien auf dem Prüfstand. oekom, München, pp 137–154
26. Chapman GP (1977) Human and environmental systems. A geographer's appraisal. Academic Press, London, p 435
27. Dietrich F (1979) Geosystemforschung und menschliches Verhalten. Geographische Zeitschrift 67(1):29–42
28. Socava VB (1974) Das Systemparadigma in der Geographie. Petermanns Geogr Mitt 118(3):161–166
29. Egner H (2006) Autopoiesis, Form und Beobachtung—moderne Systemtheorie und ihr möglicher Beitrag für eine Integration von Human- und Physiogeographie. Mitteilungen der Österreichischen Geographischen Gesellschaft 148:92–108
30. Egner H (2008) Gesellschaft, Mensch. Umwelt–beobachtet. Ein Beitrag zur Theorie der Geographie. Erdkundliches Wissen. Franz Steiner, Stuttgart, p 208
31. Roland L (2007) Kopplung, Steuerung, Differenzierung. Zur Geographie sozialer Systeme. Erdkunde, 61(2): 174–185
32. Pott A (2005) Kulturgeographie beobachtet. Probleme und Potentiale der geographischen Beobachtung von Kultur. Erdkunde, 59(2): 89–101

33. Luhmann N (1986b) Systeme verstehen Systeme. In: Luhmann N, Schorr K E (ed), Zwischen Intransparenz und Verstehen, Frankfurt, pp 72–117
34. Luhmann N (1987) Soziale Systeme. Grundriß einer allgemeinen Theorie. Suhrkamp, Frankfurt, p 674
35. Luhmann N (1992) Die Wissenschaft der Gesellschaft. Suhrkamp, Frankfurt, p 732
36. Luhmann N (1995) Soziologische Aufklärung. Die Soziologie und der Mensch, 6. Westdeutscher Verlag, Opladen, p 275
37. Luhmann N (1997a) Selbstreferentielle Systeme. In: Simon F B (ed), Lebende Systeme. Wirklichkeitskonstruktionen in der systemischen Therapie. Suhrkamp, Frankfurt
38. Luhmann N (1998) Die Gesellschaft der Gesellschaft. Suhrkamp, Frankfurt, p 1164
39. Luhmann N (2006) Einführung in die Systemtheorie, Heidelberg, p 347
40. Maturana H R, Varela F J (1984) Der Baum der Erkenntnis. Die biologischen Wurzeln des menschlichen Erkennens, p 280
41. von Bertalanffy L (1972) The history and status of general systems theory. Acad Manage J 15(4):407–426
42. Dress A, Hendrichs H, Küppers G (eds) (1986) Selbstorganisation. Die Entstehung von Ordnung in Natur und Gesellschaft. Piper, München, p 234
43. Jantsch E (1979) Die Selbstorganisation des Universums. Vom Urknall zum menschlichen Geist. Hanser Verlag, Darmstadt, p 464
44. Krüger H-P (1990) Luhmanns autopoietische Wende. eine kommunikationsorientierte Grenzbestimmung. In: Niedersen U, Pohlmann L (eds) Selbstorganisation und Determination. Selbstorganisation. Jahrbuch für Komplexität in den Natur-, Sozial- und Geisteswissenschaften. Duncker and Humblot, Berlin, pp 129–148
45. Pohlmann L, Niedersen U (1990) Dynamisches Verzweigungsverhalten bei Wachstums- und Evolutionsprozessen. In: Niedersen U, Pohlmann L (eds) Selbstorganisation und Determination. Selbstorganisation. Jahrbuch für Komplexität in den Natur-, Sozial- und Geisteswissenschaften. Duncker and Humblot, Berlin, pp 63–82
46. Jantsch E (1994) System, Systemtheorie. In: Seiffert H, Radnitzky G (ed), Handlexikon zur Wissenschaftstheorie, München, pp 329–338
47. Prigogine I (1985) Vom Sein zum Werden. Zeit und Komplexität in den Naturwissenschaften. Piper, München, p 304
48. Capra F (1983) Wendezeit. Bausteine für ein neues Weltbild. Scherz, Bern, p 512
49. Maturana HR (1980) Autopoiesis: reproduction, heredity and evolution. In: Zeleny M (ed) Autopoiesis. Dissipative Structures and Spontaneous Social Orders, Boulder, pp 45–79
50. Maturana HR (1982) Die Organisation des Lebendigen: eine Theorie der lebendigen Organisation. In: Maturana HR (ed) Erkennen: Die Organisation und Verkörperung von Wirklichkeit. Ausgewählte Arbeiten zur biologischen Epistemologie. Wissenschaftstheorie, Wissenschaft und Philosophie. Vieweg, Braunschweig, pp 138–156
51. Maturana HR, Varela FJ (1980) Autopoiesis and cognition. The realisation of the living. Reidel, Dordrecht, p 141
52. Maturana HR, Varela FJ (1982) Autopoietische Systeme: eine Bestimmung der lebendigen Organisation. In: Maturana HR (ed) Erkennen: Die Organisation und Verkörperung von Wirklichkeit. Ausgewählte Arbeiten zur biologischen Epistemologie. Wissenschaftstheorie, Wissenschaft und Philosophie. Vieweg, Braunschweig, pp 170–235
53. Maturana HR, Varela FJ, Uribe G (1982) Autopoiese: die Organisation lebender Systeme, ihre nähere Bestimmung und ein Modell. In: Maturana HR (ed) Erkennen: Die Organisation und Verkörperung von Wirklichkeit. Ausgewählte Arbeiten zur biologischen Epistemologie. Wissenschaftstheorie, Wissenschaft und Philosophie. Vieweg, Braunschweig, pp 157–169
54. Luhmann N (1997) Selbstreferentielle Systeme. In: Simon FB (ed) Lebende Systeme: Wirklichkeitskonstruktionen in der systemischen Therapie. Suhrkamp, Frankfurt, pp 69–77
55. von Foerster H (1960) On self-organizing systems and their environments. In: Yovits MC, Cameron S (eds) Self-organizing systems. Pergamon Press, New York, pp 31–50
56. von Foerster H (1984) Observing systems. Intersystems Publications, Seaside, p 331

57. von Foerster H (1987) Erkenntnistheorien und Selbstorganisation. In: Schmitt SJ (ed) Der Diskurs des radikalen Konstruktivismus. Suhrkamp, Frankfurt, pp 133–158
58. Jantsch E (1980) The Unifying Paradigm Behind Autopoiesis, Dissipative Structures, Hyper- and Ultacycles. In: Milan Z (ed), Autopoiesis, Dissipative Structures, and Spontaneous Social Orders. Westview Press, Boulder, pp 81–88
59. Jantsch E (1987) Erkenntnistheoretische Aspekte der Selbstorganisation natürlicher Systeme. In: Schmitt SJ (ed) Der Diskurs des Radikalen Konstruktivismus. Suhrkamp, Frankfurt, pp 159–191
60. Prigogine I (1967) Introduction to thermodynamics of irreversible processes. Interscience, New York, p 147
61. Prigogine I (1973) Irreversibility as a symmetry-breaking process. Nature 246:67–71
62. Prigogine I (1977) Time, structure and fluctuations (the nobel prize in chemistry 1977). In: Frängsmyr T, Forsén S (eds) Nobel lectures, chemistry 1971–1980. World Scientific Publishing, Singapore, pp 263–285
63. Prigogine I, Stengers I (1981) Dialog mit der Natur. Neue Wege naturwissenschaftlichen Denkens. Piper München, Zürich, p 314
64. Prigogine I, Stengers I (1990) Entwicklung und Irreversibilität. In: Niedersen U, Pohlmann L, (eds) Selbstorganisation und Determination. Selbstorganisation. Jahrbuch für Komplexität in den Natur-, Sozial- und Geisteswissenschaften. Duncker and Humblot, Berlin, pp 3–18
65. Prigogine I, Stengers I (1993) Das Paradox der Zeit. Zeit, Chaos und Quanten. Piper München, Zürich, p 338
66. Foerster H von (1981): Observing systems, Seaside
67. Ritter DF, Kochel RC, Miller JR (1995) Process geomorphology. Brown, Dubuque, p 546
68. Jorgensen S E, Svirezhev Y M (2004) Towards a thermodynamic theory for ecological systems, pp 354
69. Rhoads B L, Thorn C E (ed) (1996) The Scientific Nature of Geomorphology. Proceedings of the 27th Binghamton Symposium in Geomorphology held 27–29 Sept 1996, Chichester, New York, Brisbane, Toronto, Singapore, pp 481
70. Hard G (2003) Die Methodologie und die "eigentliche Arbeit". Über Nutzen und Nachteil der Wissenschaftstheorie für die Geographische Forschungspraxis. In: Hard G (ed) Dimensionen geographischen Denkens. Aufsätze zur Theorie der Geographie, Osnabrück, pp 39–66
71. Chorley RJ (1978) Bases for theory in geomorphology. In: Embleton C, Brunsden D, Jones DKC (eds) Geomorphology: present problems and future prospects. Oxford University Press, Oxford, pp 1–13
72. Hard G (1978) Noch einmal: Die Zukunft der Physischen Geographien. Zu Ulrich Eisels Demontage eines Vorschlags. Geographische Zeitschrift, 66: 1–23 (hier aus: Hard G (2003): Dimensionen geographischen Denkens. Aufsätze zur Theorie der Geographie, Bd. 2, Osnabrück, S. 113–132)
73. Hard G (1987) Die Störche und die Kinder, die Orchideen und die Sonne. de Gruyter, Berlin, New York, 22 (hier aus: Hard, G. (2003). Die Störche und die Kinder. Die Orchideen und die Sonne. In: Hard G.: Dimensionen geographischen Denkens. Aufsätze zur Theorie der Geographie. Osnabrücker Studien zur Geographie 23: 315–327
74. Goudie AS (1996) Geomorphological hotspots and global warming. Interdisc Sci Rev 21:253–259
75. Goudie AS (2001) Applied geomorphology: an introduction. Zeitschrift für Geomorphologie, supplement band (1224):101–110
76. Goudie AS (2002) Aesthetics and relevance in geomorphological outreach. Geomorphology 47: 245–249
77. von Elverfeldt K, Keiler M (2008) Offene Systeme und ihre Umwelt—Systemperspektiven in der Geomorphologie. In: Egner H, Ratter BMW, Dikau R (eds) Umwelt als System—System als Umwelt? Systemtheorien auf dem Prüfstand. Oekom, München, pp 75–102
78. Ratter BMW, Thomas T (2008) Komplexität—oder was bedeuten die Pfeile zwischen den Kästchen? In: Egner H, Ratter BMW, Dikau R (eds) Umwelt als System—System als Umwelt? Systemtheorien auf dem Prüfstand. oekom, München, pp 23–38

Chapter 2
Observation and Distinction: The Underlying Method

Draw a distinction and a universe comes into being.
George Spencer-Brown 1997

Observation has a reviving influence on science [1], and is, at the same time, basis of all knowledge and cognition. Knowledge (Latin/Greek: 'having seen') refers to visual perception—when we have seen something, we know of it (cf. Fischer in his preface to Spencer-Brown [2, p. 7]). If, particularly, all science is based on observation and if science is proceeding through it (cf. e.g. [1, p. 10]), this term has to be clarified. Penck understands specific geographical observation as 'being in the field', and thorough and accurate observation of what the geographer sees there, unveils, as it were, the problems of his subject and prepossesses him with special ideas [1, p. 9f]. With these statements Penck expresses the former—and partly also contemporary—popular opinion within geomorphology that theory automatically reveals itself just by contemplating the landscape—a literature review is, from this point of view, anything but geographic work. (also cf. Chap. 1).

However, exactly this last sentence is also a geographical observation, an observation by a geographer on how geographers work. This observation, however, focuses the research practice, the empirical studies as such, and allows us to assess the foundation of the empiricism. The question is how to comprehend these two levels or types of observation, that is the observation of the research objects (the earth surface and its forming processes) on the one hand, and the observation of the research practice on the other hand?

The cyberneticist and mathematician Heinz von Foerster (1911–2002) offers a possible answer to this question with his collection of essays "Observing systems" [3]. Heinz von Foerster, Austrian by birth, who went to the US for a dinner and stayed for a lifetime (cf. [4, p. X]), coined the aphorism "We do not see that we do not see" [4, p. 26], which already expresses a central aspect of his observation theory. Within the theoretical context of Heinz von Foerster the term observation goes further than in our everyday understanding. On the one hand, all systems[1] are capable of observing,

[1] However, Foerster understands systems differently, namely as being self-referential (also cf. Chap. 3.1).

K. von Elverfeldt, *System Theory in Geomorphology*,
Springer Theses, DOI: 10.1007/978-94-007-2822-6_2,

be it biological, psychological, or social systems. Moreover, the theory of observation results in narrow limits for objectivity, and it forms one of the pillars of radical constructivism.

Heinz von Foerster gained his insights from then pioneering studies on neurophysiology and on the cognitive ability of living beings. The basic statement is that every observer has a blind spot, that is, something he or she does not perceive. To start with, this can be very well understood as the physiological blind spot of the human eye. Peculiar about this is, however, not the fact that something is not seen, but rather the fact that the observer is not capable of perceiving that he does not see (something). To put it more simply: We do not permanently have a black spot within our visual field as our brain fills the gap in a quasi-meaningful way (also cf. [5, p. 21ff]). This means that we are blind for the blind spot, and in this sense every observation is blind (cf. [6, p. 38ff]). This can be illustrated by studies on changes in behaviour after certain brain injuries (cf. [3, 4]). For example, certain cerebral injuries result in a considerable loss of the visual field of which, however, the person affected is unaware. Though, certain motoric dysfunctions develop such as a one-sided loss of control of arm or leg. One possible therapy is to blindfold the patient so that he learns to pay attention to the 'normally' working inner channels that signal him his body posture. What is important for us here, is that the missing perception is *not* perceived, but that this perception has to be trained with senso-motoric interaction. What is being perceived thus depends on the inner structures of an organism and is not triggered by external signals.

However, it is not just the visual perception that exhibits blind spots, but for example also hearing. Foerster [4, 5] mentions an experiment during which a tape with one and the same word is constantly repeated. After a specific number of iterations the probands start to hear something else, so called "alternants". Another example refers to animal experiments where the internal construction of perception could also be shown: A cat, prepared with the respective micro-electrodes, is put into a cage within which it only reaches its food if it pushes a lever just when a certain sound occurs. That is, the cat has to learn that a certain sound means 'food'. The recorded patterns of its neural activity show that the cat does not perceive the sound—until it realizes the link between sound and food. In other words: only if a perception becomes comprehensible the whole (neural) system starts to work (cf. [4, p. 28–29]).

According to the mathematician and Nobel laureate Bertrand Russell (1872–1970), such physiological findings lead the naïve realism ad absurdum [7, p. 127] (cf. Chap. 9), as these examples show that something is seen or heard effectively is not 'there' (the alternants), or that though it is not seen or heard, something is 'there' (the limited visual field or the 'sound for food'). These phenomena cannot be explained by a naïve-realistic position as they contradict the conviction that our sensory organs picture the 'world as it is'. This, however, is a conviction which is reflected within geomorphology in the assumption that after sufficient training we can read within the landscape like in a book. Especially the cat experiment shows, though, that we already need a theory about how things are connected (food and sound) in order to actually perceive these connections: Observation without theory is impossible.

Foerster corroborates his observations by introducing the "principle of undifferentiated encoding":

> The response of a nerve cell does not encode the physical nature of the agents that caused its response. Encoded is only 'how much' at this point on my body, but not 'what' [4, p. 29].

The attribution of the stimulations to specific sensory receptors only takes place in the brain as any sensory organ is (normally 'correctly') coupled to a specific brain area (also cf. [8]). Hence, our brain works in a very similar manner to an engineer at his control console: A red light in this column and that row assures him that exactly such and such a machine is defect [8]. This then means that all our sensory receptors are 'blind' to the *quality* of the stimulation as they are just reacting to its *quantity*. Even if we find this surprising it should not amaze us: 'out there' indeed is neither light nor colour, but only electromagnetic waves; 'out there' is neither sound nor music, just molecules that are moving with more or less mean kinetic energy etc. [4, p. 29].

On a more theoretical level this aspect of a blind spot—which so far has been dealt with only on physiological or experimental grounds—can be summarized with the words of the mathematician George Spencer-Brown (1997): Draw a distinction. After all, the

> act of indicating any being, object, thing or unity involves making an act of distinction which distinguishes what has been indicated as separate from its background. Each time we refer to anything explicitly or implicitly, we are specifying a criterion of distinction, which indicates what we are talking about and specifies its properties as being, unity or object [5, p. 46].

This is a thoroughly ordinary and not a special situation in which we inevitably and continuously find ourselves. Observation thus is defined as the twofold practice of distinction and simultaneous indication of one side of what we have distinguished afore (vgl. [9, p. 69]). This distinction and indication of something, which necessarily leaves all other things aside, is the starting point of every observation and enables us to establish a network of distinctions. Such, information on the observed can be gained (cf. [10, p. 125])

According to this understanding, the act of observation is not exclusively human anymore as it has been suggested for centuries by the western tradition of thought, since for example also an amoeba is capable of observation in this sense—otherwise it would possibly devour itself (cf. [12]).

Thus, observation is done actively—"perceiving means acting"[4, p. 27] – since the observed is distinguished from all other possible things and is indicated as something distinct, e.g. 'river', 'landslide', and 'talus slope'. That is, 'something', a thing or an object only becomes visible through this act of distinction and indication. In the very moment at which we observe (distinguish and indicate), everything else takes a backseat, we do not regard it. For this reason the credo of the observation theory is "observe the observer", that is, the self-observation as well as the observation of the world is shifted from first order observation (asking "What is being observed?") to the level of second order observation, asking "How

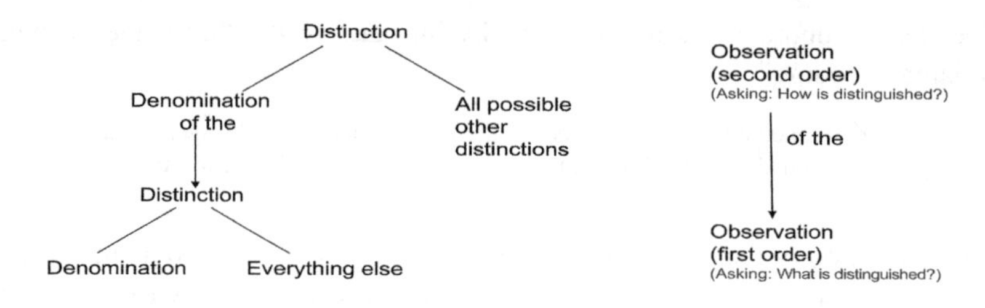

Fig. 2.1 Schematic sketch of the observation theory and of the difference of first and second order observation ([11], modified)

is something being observed?" (cf. Fig. 2.1). It is only at this level that the respective blind spot of the observer can be realized: the blind spot is the respective underlying distinction, which guides what is being observed and what is not being observed. Second order observation is the observation of the observation and can thus reveal the distinctions the first order observation is based upon. Hence, while first order observation brings forth an object, second order observation brings forth the acting, the basic distinction. Furthermore, second order observation usually utilizes two distinctions: First, it distinguishes between the observer and his/her object, second, it distinguishes which kind of distinction the observed observer is using (cf. [12, p. 46f]). From this it becomes clear, however, that with each observation there will result a new blind spot: Any second, third, fourth … order observation is, at the same time, also a first order observation and therefore blind for something else. Consequently, a fixed reference point for observation from which everything can be impartially and objectively observed, a so-called *locus observandi*, cannot exist. Thus, there are also no grounds on which to judge better or worse, wrong or right, which certainly is a thought that shakes the very foundations of scientific thinking. It points towards a thinking according to which there is no ultimately true observation of the world 'as it is', as from this perspective everything is dependent on the observation and the distinction which have been made. If the above is taken seriously, then it becomes clear that the descriptions of reality are endlessly recurring and finally vanish into thin air, just as any clearness and validity. Against this background observers can decide to take over a—paradox—self-foundation (cf. [13]): The decision to support this and no other statement as being valid here and now. Keeping this paradox foundation in mind, one can stay capable of learning, as—after all—within this theoretical perspective statements of 'real' truth and rightness or falsity are impossible. What can be shown, however, is which descriptions, results, theories are now viable[2] and with which we can now work effectively.

[2] The term viability was coined by Ernst von Glasersfeld (e.g. [14, 15]) and means the functioning of ideas. It is a constricted concept of 'truth', as according to the observation theory it is impossible to ultimately state whether a theory is true or false (also cf. Chap. 9.1).

When these thoughts are applied to geomorphology (or other scientific disciplines), it becomes apparent that, first, our observations of the world are solely based upon the internal structures of the cognitive apparatus, and that, second, these observations are largely based upon specific, yet unconscious presuppositions ('distinctions'). For example, as geomorphologists we 'automatically' observe the landscape according to specific criteria of distinction such as form, process, matter, scale, whilst we do not—or only in the second instance—apply other, equally possible criteria such as 'energy budget'. These initial criteria of distinction that are substantially influenced by the actual paradigm form the directives for the decision about which further research is possible and even feasible. Ute Wardenga [16], a German geographer, named this trained point of view, which is specific for every scientific discipline as "Wahrnehmungsdressur", that is cognitive dressage. Once more, from this it follows that observation without theory is impossible (also cf. [17, p. 51]).

That observations are theory-dependent can be illustrated by the scientific use of instruments that are generally seen as being more objective than our own observations. But what is an instrument? After all, it is nothing more than an extension or an appendix of our own sensory organs, but which has three tasks to fulfil: First, to make accessible those areas which are normally not directly accessible via our senses, second, to make measurements within these areas, and third, to transform these measurements in such a way that we can perceive them (cf. [18, p. 127]). Therefore, measuring instruments are no "neutral observers", quite the contrary, as they serve a purpose, namely to confirm a theory. Already the construction of the instrument needs a lot of theory: instruments are theory-loaded. In my opinion, one of the best examples is time, or the watch: According to the theory of the big bang, time came into existence together with matter, energy and space at moment of the big bang [19]. Ever since, the brightest minds have pondered the question what 'is' time, after all, and whether it 'exists'at all, but so far no answer has been found. Time is a purely theoretical concept (cf. e.g. [20]). Thus, although we don't even quite know what time is, we do have the most exact instruments to measure it.

One can even take the conclusion further that observation without theory is impossible if one states that there is no observation which does not act back upon the observed procedure.[3] Erich Jantsch (1929–1930) [21], an Austrian nuclear physicist and co-founder of the Club of Rome states in his monograph "Die Selbstorganisation des Universums"('the self-organisation of the universe') that it

[3] For example, for the observation of elementary particles the choice of the measurement—that is, the type of observation—technique is crucial as this determines whether such a particle is perceived as wave or matter (cf. Chap. 9.2): Hence, the observation acts back on the observed. Generally, such an influence is also accepted for macroscopic systems, but is seen as neglectable. An example from geomorphology could be BTS-measurements during which inevitably a heat flow is initiated but this influence is seen as being too minor. As we will see in Chap. 5, such small changes or fluctuations can theoretically have a major impact if the system is beyond an instabitity threshold. Presumably, Jantsch is referring to such issues.

is rather ironical that the influence of the observer on the observed process was first formulated for the subatomic field (the so-called Heisenberg uncertainty principle). Jantsch compares the observation of subatomic processes with a watchmaker who uses a bulky hammer to get to grips with a ladies' watch. According to Jantsch this is ironical, because the influence of the observer is much more obvious in macroscopic systems as every action, every thought, and every theory interferes with our research object [22, p. 54].

These remarks show why it is necessary to specify conditions for any statement under which it gains clarity and validity. The decision for or against the observation of something can only be seen in retrospect and/or by another observer, that is by means of second order observation. This is also (or even especially!) true for this thesis: The criteria under which it is valid are the distinctions upon which it is based. After all, these distinctions determine what can be observed: For example, with the difference of good or bad I can—no matter what I am looking at—see something else than with the difference of rich and poor, beautiful and ugly, new and old, or healthy and sick [23, p. 34f].

Within this thesis I have utilized Foerster's observation theory and its methods in order to observe the (implicit and explicit) theoretical basis and its application within geomorphology, and to examine the epistemological and practical consequences that are attached to it. Here, I used the distinctions 'system theory within geomorphology'and 'system theories in other disciplines', or rather 'first order system theory' and 'second order system theory'. Following, the distinctions have been '*coherency and stringency*' and '*incoherency and logical breaks*', as well as '*high connectivity (to other disciplines)*' and '*low connectivity*'. Hence, I was only capable to observe what I have observed—the relevance or irrelevance of geomorphology for our society, for example, cannot be part of my examinations. Therefore, I cannot—and do not even want to—claim any truth (whatever that might be) or picture any geomorphological research 'reality', as this is simply impossible if the theory of observation is taken seriously. It is nothing more or less than *one* way to view geomorphological scientific practice. However, I believe that only if we as geomorphologists perceive our research (objects), that is, based on which distinctions, we will be able to understand which specific positions and with which specific presuppositions we are conducting our studies. Furthermore, we can then actively direct our attention towards those problem areas that we have not perceived before and thus make them visible. If we re-think Penck's words in this way we indeed might state: Observation has a revitalizing influence on science!

References

1. Penck A (1906) Beobachtung als Grundlage der Geographie. Borntraeger, Berlin, p 63
2. Spencer-Brown G (1996) Wahrscheinlichkeit und Wissenschaft. Carl Auer, Heidelberg, p 142
3. von Foerster H (1984) Observing systems. Intersystems Publications, Seaside, p 331

4. von Foerster H (2006) Sicht und Einsicht. Versuche zu einer operativen Erkenntnistheorie. Carl-Auer, Heidelberg, p 233
5. Maturana HR, Varela FJ (1984) Der Baum der Erkenntnis. Die biologischen Wurzeln des menschlichen Erkennens, p 280
6. von Foerster H (2002) In jedem Augenblick kann ich entscheiden, wer ich bin. In: Pörksen B (ed) Die Gewissheit der Ungewissheit. Gespräche zum Konstruktivismus. Carl Auer, Heidelberg, pp 19–45
7. Russell B (1952a) Die letzten Bestandteile der Materie. In: Russell B (ed) Mystik und Logik. Philosophische Essays. Humboldt, Wien, pp 125–144
8. Roth G (1987a) Erkenntnis und Realität: Das reale Gehirn und seine Wirklichkeit. In: Schmitt S (ed) Der Diskurs des Radikalen Konstruktivismus. Suhrkamp, Frankfurt/Main, pp 229–255
9. Luhmann N (1998) Die Gesellschaft der Gesellschaft. Suhrkamp, Frankfurt/Main, p 1164
10. Baraldi C, Corsi G, Esposito E (1997) GLU Glossar zu Niklas Luhmanns Theorie sozialer Systeme. Suhrkamp Taschenbuch Wissenschaft, Frankfurt a. M., p 248
11. Egner H (2008) Gesellschaft, Mensch Umwelt—beobachtet. Ein Beitrag zur Theorie der Geographie. Erdkundliches Wissen Franz Steiner. Stuttgart, p 208
12. Fuchs P (1992) Niklas Luhmann—beobachtet. Eine Einführung in die Systemtheorie. Westdeutscher Verlag, Opladen, p 219
13. Bardmann T(2001) Eigenschaftslosigkeit als Eigenschaft. Soziale Arbeit im Lichte der Kybernetik des Heinz von Foerster. Online Journal für systemisches Denken: Das gepfefferte Ferkel pp 1–13
14. von Glasersfeld E (1987) Siegener Gespräche über Radikalen Konstruktivismus. In: Schmitt SJ (ed.) Der Diskurs des Radikalen Konstruktivismus. Suhrkamp, Frankfurt/Main, pp 401–440
15. von Glasersfeld E (1992) Why i consider myself a cybernetician. Cybern hum knowing 1(1):21–25
16. Wardenga U (2001) Zur Konstruktion von Raum und Politik in der Geographie des 20. Jahrhunderts. In: Reuber P, Wolkersdorfer G (eds) Politische Geographie: Handlungsorientierte Ansätze und Critical Geopolitics. Heidelberger Geographische Arbeiten, Heidelberg, pp 17–31
17. Rhoads BL, Thorn CE (eds) (1996) The scientific nature of geomorphology. In: Proceedings of the 27th Binghamton symposium in geomorphology held 27–29 September 1996, Chichester, New York, Brisbane, Toronto, Singapore p 481
18. Peschl MF (2001) Constructivism, cognition, and science—an investigation of its links and possible shortcomings. Found Sci 6(1–3):125–161
19. Dinwiddie R (2003) Die Erdgeschichte. In: James Luhr F (ed) Die Erde. Doring Kindersley, London, pp 20–41
20. Falk D (2008) In search of time: journeys along a curious dimensions. McClelland and Stewart, Toronto, p 344
21. Jantsch E (1980) The Unifying Paradigm Behind Autopoiesis, Dissipative Structures, Hyper- and Ultacycles. In: Zeleny M (ed) Autopoiesis, dissipative structures and spontaneous social orders. Westview Press, Boulder, pp 81–88
22. Jantsch E (1979) Die Selbstorganisation des Universums. Vom Urknall zum menschlichen Geist. Hanser Verlag, Darmstadt, p 464
23. Pörksen B (2002) Die Gewissheit der Ungewissheit. Gespräche zum Konstruktivismus. Reihe Konstruktivismus und systemisches Denken. Carl-Auer-Systeme Verlag, Heidelberg, p 237

Part I
Problem Areas

Chapter 3
First Problem Area: Coherence of Basic Assumptions and Concepts

The beginning of wisdom is the definition of terms.
Socrates

The term 'system' is largely accepted as interpretation pattern within geomorphology, which is reflected in the amount of publications within which system theory serves as theoretical reference point. A search within the ISI Web of Science points to a strong increase of geomorphological system theoretical research within the 1990s and the first decade of the twenty first century.[1] From 1960 to 1989 the amount of publications that referred to "geomorph" and "system"[2] was only 28 of 903 (<5%). Within the following decades, however, system-theoretical studies showed an increase in numbers: approx. one-third of all geomorphological papers showed reference to systems in some form [27% (971 of 3,656) of the publications within the 1990s and 31% (2,205 of 7,044) from 2000 to 2009]. Although the reliability of these numbers is limited and, consequently, they are not supposed to stimulate any further analyses, it can be shown on a random basis that the theoretical foundation as well as the definitions and basic assumptions are rarely, if at all, reflected and analysed. This can be seen as an indication that systems are seen as given and 'natural' or obvious.

Thus, geomorphological system theory clearly shows paradigmatic traits (cf. [1]) and implicitly determines research foci. Additionally, studies generally (and maybe already traditionally) focus on empirical work, which could partially explain this attitude. Theoretical discourse does not rate high within geomorphology:

> [...] geomorphological research and the literature it generates remain dominated by empirical case studies [2, p. 47].

However, to be clear: With this it is by no means stated that empirical studies are of low scientific value, or even that there has not been any progress in the

---

[1] This search does not claim completeness and only serves as an indicator for a development. Furthermore, only those articles were captured that utilize "system" within title, abstract, or keywords.

[2] The search algorithm was 'geomorph*' AND 'system', and for the determination of the reference frame 'geomorph', respectively.

K. von Elverfeldt, *System Theory in Geomorphology*,
Springer Theses, DOI: 10.1007/978-94-007-2822-6_3,

system-theoretical foundation of geomorphology. The very existence of studies on non-linear, complex, chaotic and/or self-organized system behaviour (e.g. cf. [3, 4, 5, 6–11]) already shows the progress. However, this does not change the observation that there exists a considerable disequilibrium within geomorphology of empirical towards theoretical studies.

These two developments—the matter of course of the system theoretical perspective as well as the "epistemological imperative of empiricism" (Richard Dikau 2008, personal communication)—are significantly hampering the theoretical foundation and the further development of our discipline. Questions on *why* we consider nearly all phenomena as system, or whether there are some restrictions to system-theoretical analyses, or even what is a system in stricter sense are—if at all—only rarely posed.

According to Kuhn [12] this development can be seen as a classical characteristic of a paradigm within the phase of normal science. However, the question is: Can a scientific discipline that traditionally has such a strong focus on empirical work be explained by Kuhn's concept of changing primarily empirical and theoretical analysis of study objects? Otherwise this could also be seen as an example of a scientific theory to an implicit theory[3] (cf. [16]). This perspective offers an advantage: An implicit theory, that means a conglomerate of assumptions that are no longer made explicit and reconsidered, can be re-transferred to a scientific theory. For this it is paramount to review and to explicate basic assumptions and concepts. For example, wide parts of geomorphological research seem to be unaware that systems theory is based on specific basic assumptions that *may* apply to geomorphological questions, but *need not do so* (cf. [17, p. 213]).

This geomorphological research practice that gives high priority to empiricism but not to theory virtually forms the background for all problem areas which will arise in the following.

3.1 On Unity, Interaction and Boundaries[4]

> But what are the ideas of the laws of form? They are probably more easily told than understood. The simplest instruction is to draw a distinction. By obeying this order two states are created, a marked state and an unmarked state, a double entendre which contains a trivalence: "What is the thing, what is it not, and the boundary in-between".
>
> (Rudolf Maresch on the "Laws of Form" of George Spencer-Brown)

[3] The term "implicit theory" stems from psychology where it is understood as individual constructions by single persons of specific phenomena [13, 14]. This concept shows some similarities to the concept of "crypto theories", which has been coined by Peter Weichhart within (German) human geography and which has its roots within literary studies (cf. [15]).

[4] Within this chapter thoughts are united which have only been published independently so far: Elverfeldt and Keiler [18], Egner and Elverfeldt [13] und Elverfeldt and Glade [19].

Within the preceding section it has been shown that geomorphology deals with theory mostly implicitly. Now it is important to explicate these implicit assumptions and thus to make them accessible for discussion. Definitions of the objects of studies are some of the most essential constituents of theories. Therefore, in the first instance rather simple, if not banal questions form the guideline for the next sections: What is a system? As well as: How can geomorphic systems be characterized?

An obvious approach to the question what is a system is to consult diverse lexica. The 'Handlexikon zur Wissenschaftstheorie' ([20, p. 329], translation by KvE) defines system as follows:

> Greek sýstēma: the compounded [...] In older definitions, especially those from the field of engineering, *system* is understood as a variety of components that start to relate to each other. A more modern view, which gains ground only lately, in contrast stresses that processes are forming a network of relations. This expresses a change from a traditional, spatial structure-focussed thinking towards a modern (and at the same time ancient) process thinking.

If this distinction is followed it can be observed that within geomorphology the older, technical definition of systems is dominant, which understands systems as a variety of interacting components (e.g. cf. [21, p. 225]), just as it is expressed within the 'Lexikon der Geographie' ([22, p. 24], translation by KvE):

> 1) Generally: Composition, arrangement, consistently ordered whole; a central concept of the system theoretical natural and social sciences with an important origin within the functional approach of biology. System is a model-like representation of reconstructed relations between different elements of a whole. Thereby it is assumed that any arbitrarily thematically/spatio-temporally delimited observation unit can be understood as system [...].

This picture is matched by the definition presented by Richard Chorley and Barbara Kennedy [23, p. 1]) within their fundamental opus on geomorphological system theory:

> A system is a structured set of objects and/or attributes [...] that exhibit discernible relationships with one another and operate together as a complex whole.

This means that within geomorphology in the widest sense any research object can be addressed as system, which forms a unity and shows interdependencies of its elements (also cf. [24, p. 303]).

Despite the fact that the geomorphological system theory has mostly only been applied implicitly (that is, without the need of definition), the term system is actually defined in some recent publications. However, these definitions mostly do not differ from that which Chorley and Kennedy introduced to geomorphology in 1971 (e.g. cf. [25, 26]). For example, Phillips states that systems are structured unities, which are defined by us as observers:

> A system is a set of interconnected parts which function as a complex whole [5, p. 195].

However, Phillips does not explain what exactly is a "complex whole". The problem is—amongst others—that the difference between system and 'everything else' is not made clear.

Thus, these definitions show that the idea that systems build a 'unity' or a 'whole' is the starting point of thinking within geography—and especially within geomorphology. 'Unity' and 'whole' imply a boundary to an 'outside' that makes a system discernable as being distinct. For this, the importance of the boundary should not be underestimated, since

> [t]he boundary has special significance in systems theory as it, by definition, separates the system from its environment. In so doing, it effectively defines and operationalizes the system. The significance of the boundary for both intra-system relations and inter-system relations cannot be overstated [27, p. 1].

However, the above definitions are rather fuzzy as to how this boundary becomes discernible. In other words: What (or even who) turns a system into a system? Even though Chorley and Kennedy [23, p. 23] answer this question by referring to common sense (sic!) and the experience of the scientist, they finally state that it is surprisingly difficult to delimit even the most simple imaginable system (i.e. morphological system). The problem is that for system analysis one needs to know and to describe all elements that are part of the system, as well as the relations between these elements. Hence, all elements and all the interactions between them would have to be known beforehand in order to describe a system—that is, the result of the analysis is at the same time the starting point of the analysis. Possibly, phenomena that might be essential are thereby excluded (also cf. [28]). Thus, the seemingly simple definitions above tautologically refer to themselves instead of upon any demarcation criteria, that is, something that is *not* the system. This way, a system already has to have been analysed in order to be analysed and defined. The definitions are circular. The problems that are attached to this could possibly be circumvented, if the definition and the following analysis were used as hypothesis with the aim of falsification or verification. However, this does not seem to be the case.

In order to delimit a specific system the scientist faces a difficult task, as he or she needs to define a set of variables or criteria that makes a system distinct from its environment. As, additionally, any "arbitrarily thematically/spatio-temporally delimited observation unit can be understood as system", as Brunotte et al. ([22, p. 324], translation by KvE) put it, any specific system has to be re-defined and re-delimited according to the respective research question. That is, anything that is defined as system in one study can either be only a side-aspect, or an element, or just a variable within the next study.

Generally, it is of major importance to specify adequate and correct limits for the definition and delimitation of systems as otherwise the results might be paradox and absurd (cf. [29, p. 2]). However, this is extremely difficult to achieve with the definitions above. Hence, in most cases physical features are utilized, which can be easily recognised or interpreted as boundaries, e.g. the boundary between liquid/solid (as for the boundary between a slope and a river), frozen/unfrozen (as for the delimitation of a permafrost body) (e.g. cf. [26, 30], and for a critique on such a method [31]). For example, White et al. ([30], p. 11], original accentuation) state:

> [A] [...] system is confined to a definite place in space by the *boundary* of the system, whether this is natural and real like a cell wall or watershed, or arbitrary, though still real, like the walls of a test-tube or vessel in a laboratory, or arbitrary and intangible like the boundary of a cloud.[5]

Furthermore, the scientist also determines the type and amount of criteria, the 'success' of which highly depends on his expertise, but also on his creativity, that is, his ability to analyse those variables that he has neglected so far whether they are relevant for the system under study. Consequently, the delimitation of systems is to a certain extent purely subjective and arbitrary, instead of being guided by a consistent principle (also cf. [28]).

The thought that a system is the smallest network of all of those elements that are non-linearly coupled to each other and that only show minor, linear interactions across the system boundaries is another, somewhat more precise answer on how to define and delimit a system [32, p. 395]. Probably, Robert W. Christopherson aims in a similar direction when suggesting that the internal energy and mass flows are different from the external ones and that this difference can be used to delimit systems from their environment [33, p. 4]. The advantage of these two definitions is that they implicitly and explicitly refer to an outside, thus offering criteria for the distinction of a system from its surrounding. However, Werner and McNamara [32] do not answer the question how to measure the 'strength' or the linearity of interactions between elements, and whether this is a meaningful distinction of system and environment, respectively. Also Christopherson [33] does not explain in detail the distinguishing feature of system internal flows so that they become discernable. To start with it would be useful to know how to define 'the inside'—a problem that arises

> Whenever we have to deal with systems which do not come wrapped in a skin. In such cases, it is up to us to define the closed boundary of our system [34, p. 119].

Thus, within geomorphology systems are delimited by the observer as he analyses the specific network of elements and their interactions and their relations to the environment. Most geomorphologists probably agree with the chemist Jurgen M. Honig who defines systems as that part of the universe that is set apart from the remaining cosmos for special study [29, p. 2].

Thus, it becomes clear why a definition does not seem to be essential: Any study object can be described as system. The boundaries are given by the research focus, as a system simply is the respective section under study (cf. [35, p. 2], cited after [36]). The characteristics of the system and how it thus behaves no longer is an object property, but solely depends on the location, more correctly: the choice of the system boundary [37]. Figure 3.1 graphically depicts this aspect: the dashed line that represents the system boundary and that takes different positions in Fig. 3.1a than in 3.1b exemplifies how strongly the choice of system boundary influences the analysis. For example, in- and output relations are changing with the

---

[5] At the same time, this seems to be a good example forthe double ontological non-sequitur (doppelter ontologischer Kurzschluss) mentioned by Hard.

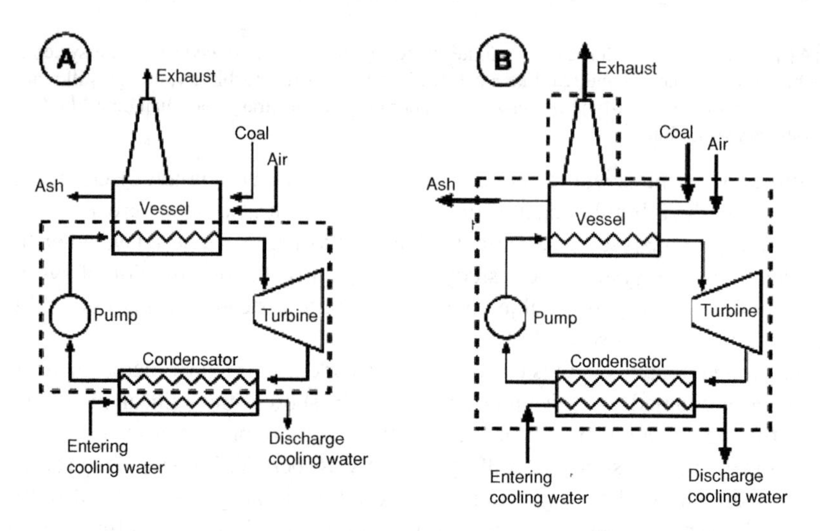

Fig. 3.1 The system character depends on the choice of system boundary (Source: Sattelmayer [37], modified)

location of the system boundary, and by omission or addition, respectively, of different system elements the focus changes considerably. Hence, two studies that 'originally' share the same object of study achieve results that are not comparable. Additionally, also the prediction of system states is highly dependent on the choice of system boundary. As there is no coherent rule for the delimitation of systems it cannot be known which system analyses is 'correct', this arbitrariness of delimitation sufficiently contributes to the perception that the world is too complex for decisions (also cf. Chap. 4).[6]

The above definitions show that the concepts of unity, elements, and interactions are characteristic for systems thinking (cf. [39, 40]). Systems are an interrelationship of single elements, a unity, or a context. In this understanding, unity emerges from the elements and the interactions between them. The basic idea is that the current system state can be completely described by the internal composition and the framework conditions (also cf. [41, p. 13]).

From this short overview one of the core problems of the geomorphological systems theory becomes obvious: The non-definition of systems, which *de facto* can be found in nearly all studies. From this point of view, everything is a system, and thus the concept of systems is applicable to literally anything—however, something that cannot be wrong also discloses nothing ([42, p. 30], also cf. [43, 44]). Systems, as they are traditionally seen within geomorphology, do not have definite boundaries, and as a result the term' system' is not much more than a synonym for 'context' or 'correlation'.

[6] With Fuchs [38] I will call this the exculpation function ("Entschuldigungsfunktion") of first order system theories.

This is opposed by the view that a system is more than a context or correlation, and that one should only name something 'system' when a context or correlation delimits itself from an environment (cf. [45, p. 17]). Based on the theory of observation of Heinz von Foerster ([34, 46] also cf. Chap. 2) and the works of George Spencer-Brown [47] a different approach to defining systems results. According to them unity cannot be the starting point for observation, but only a difference, in this case the difference of system and environment (also cf. [48]). It all starts with the assumption that something (anything!) can only be perceived if it is discernable from something else (cf. Chap. 2). In the case of systems this means that the system and its environment are different, and it is this difference which allows delimiting the system. Within his calculus "Laws of form" Spencer-Brown [47, p. 1] states that a distinction is needed whenever one side and not the other is to be denominated—any denomination is impossible without distinction. In this sense a distinction is a boundary, the demarcation of a difference which leaves the environment outside. The system is on one side, the environment on the other [49, pp. 74–75].

When a space is separated in two parts by a distinction (e.g. a circle or a boundary line) two sides of the distinction come into being, an inside and an outside. The inside and the outside, or to put it differently: the defined and the undefined have an identical form (cf. [47, p. 9]). The form therefore denotes the unity of two sides: A distinction, for example made by a definition of 'this', is at the same time the delimitation or the description of 'this', and also of everything else except 'this' (cf. [50, p. 64]). As the denoted (the system) and the non-denoted ('what it is not'—the environment) are of the same form, George Spencer-Brown [47] speaks of the unity of the distinction of the denoted (the marked) and of the non-denoted (the unmarked).

In this perspective thus the distinction builds the starting point for thinking, and not a unity of a system that somehow emerges from elements and interactions. But how can such a difference between system and environment be recognized? One viable approach for the delimitation and definition of systems is offered by the concepts of autopoiesis and self-reference (cf. [49, 51–54]). The concept of autopoiesis (self-creation) describes the ability of living systems[7] to produce and reproduce all elements of which they are built [54, p. 50] whereby they maintain themselves.[8] In the case of living systems the autopoiesis, i.e. the *modus operandi*

---

[7] In their early works, Maturana and Varela [55] describe living systems as machines. This was indeed meant as provocation and especially as alternative approach to animistic tendencies (also cf. [56] and Chap. 1). Other authors also describe living systems as machines (cf. [57]).

[8] Maturana [53] call those systems that do not reproduce themselves by their own operations but something different, and which thus do not set their own boundaries allopoietic. An allopoietic system is a product of another, system-external process, and as it thus has to be defined by an external observer, it is a purely analytical construct [58]. Maturana [53] state that autopoietic systems can be equally studied allopoietically, that is, with regard to their input–output-relations. However, this point of view limits the degree of potential insights on the inner organisation. It is the combined approach that leads to a recognition of the system's relations to its environment (also cf. [56] and [58]).

that can only be found within living systems, is cell division, that is, a continuous reproduction of cells by cells. In other words this means that any cell of a living system has been produced by cell division, which again is a result of a network of system internal processes, and which thus cannot be attributed to the influence of e.g. another system or the environment as such. Maturana and Varela [54] have described this special mode of operation, autopoiesis, for living systems; autopoiesis is consequently the characterizing differentiator of life and non-life. The immune system or the nervous system on the other hand are self-referential, i.e. they refer to themselves with all their operations, but they are not autopoietic since they do not produce themselves (also cf. [56]).

As this mode of operation (cell division) only occurs in living systems it serves as distinctive feature to all other possible types of systems. Hence, there is no necessity for the scientist to search for a unity of variables or for criteria that form a system. After all, all those elements that are not created by the system itself do not belong to the system—thus, the self-creation sets the limits and unity of the system. In this sense the cell is a product of internal operations and not a result of inputs and/or outputs. Any operation within the system depends on preceding operations whilst it also forms the basis for future operations. Thus, the self-production is circular or, respectively, a closed cycle (cf. [49, 59]), and thus autopoiesis is a self-referential process itself. Hence, the system refers only to itself in all its operations and not to its environment or another system. This does not exclude the possibility that a system reacts to what it is exposed to—but only in such a way as provided for by its structures [60, 61].

This perspective differs fundamentally from the traditional geomorphological perspective according to which a system (a slope, a river, or a glacier) is understood as a complex whole that exhibits clear cause- and effect-relations internally as well as externally. However, autopoiesis is not the antonym to interrelations with the environment: it is an antonym to causalities [61]. Hence, autopoiesis, as a form of self-reference, continues the thought of self-organisation, or rather exceeds it by far (also cf. [50]). (Self-)Organisation (also cf. Sects. 4.2 and 4.3) refers to the relations that have to exist between elements ([54, p. 54], also cf. [62, p. 64]), and by which any unity becomes defined as a specific unity (vgl. [63, p. 19]). These relations include self-control and self-regulation as the system utilizes self-imposed rules for how it handles itself and its environment. In this perspective single components are just studied with regard to their role for the creation of the unity that they integrate. If the *organisation is studied*, findings on the characteristics of these components are impossible, as these are part of an *analysis of the structure* of the system. For example, the specific organisation of cell division can be become manifest in different structures, e.g. wood, other plant fibres, or in animal cells.

Autopoiesis and self-reference include a self-determination which is further reaching than the thought of self-organisation: Autopoietic systems independently produce and reproduce their components and herewith create their unity. Thus, autopoiesis includes the relations as well as the components of the system, which specifically built a specific unity and materializes their organisation. For example, a living being is characterised by its autopoietic organisation (the cell division). Different living beings differ according to different structures, but they are the same as far as their organisation is concerned [54, p. 55].

Due to this self-creation and self-reference autopoietic systems are autonomous. This autonomy should not be confused with autarky: Specific environmental conditions (as enough nutrition or a specific temperature range) are required for the existence of the system (also cf. Sect. 4.1). These pre-conditions, which are—in traditional systems theory—seen as input factors and framework conditions, are thus externalised. That is, as long as the system exists it is automatically adapted, just as long as it exists. According to Maturana and Varela [54] there is no such thing as "well" or "badly" adapted which again clarifies that the influence of the environment on the system is strongly limited. In consequence, this concept has far-reaching implications for the understanding of systems: If the system creates itself and if it refers only to itself in all its actions, there is only one way in which environmental factors can determine system behaviour (also cf. Sect. 4.1), that is, if changing environmental conditions end the existence of the system.

In the perspective of the second order systems theories of biology, mathematics, and sociology it is thus fundamental to draw a distinction, a distinction of system and non-system. At the same time, this is no longer an arbitrary or an optional distinction as it follows a clear rule: It is the system itself that sets its boundaries, either by owning a specific process complex,[9] which determines the autopoiesis of the system (and which can only be found within the system and never outside), and/or by building a specific self-referential type of organisation which makes it distinct from its environment. From this point of view, systems cannot be understood as input–output-models anymore, which function according to more or less simple causalities [50]. According to second order system theories, systems are more than a section of the earth's surface, thereby performing a transformation process of energy and matter through the interaction of elements. Autopoiesis can thus be seen as a structure generating process, and the resulting structured systems act in a self-organised and self-referential manner. And it is this act of building structures which is a core interest of geomorphology: After all, it is the aim of geomorphology to explain the structures of the earth's surface—that is, landforms—and to infer from there to the building processes. However, how should this be possible with an approach which, in the end, only focuses on flows of mass and energy? These describe the metabolism of a system, but not how this metabolism is utilized for structure building. How this metabolistic system understanding of input–output-models coins geomorphology, and which limitations it brings with it will be shown in the following Sects. 3.2–4.2.

[9] The process structure that enables autopoiesis consists of the processes of interaction, production, transformation, and destruction [64].

3.2 Organising Principle: Complexity?

One of the main tasks of science is to create some order in our world (cf. [65]), so that the world becomes comprehensible and subsumable. Seemingly, geomorphology does not fully meet this task as Pitty [66, p. 66] dunned for a reduction of geomorphological complexity. Considering the previous sections this might sound paradox, as there geomorphology was criticized for having too simple an approach when applying linear input–output-relations to geomorphological issues. But in the end, this is precisely the problem: An approach of energy and mass flows that are arbitrarily defined only has small organising potential for science. As the earth constantly receives free energy (mainly from the sun) these flows proceed continuously whilst one quasi gives the impulse for the next. As a consequence, geomorphic systems appear to be endlessly complex and interconnected since outputs cause inputs and vice versa.

Ultimately, from this perspective it is obvious to differentiate systems according their degree of complexity. Such an organising principle is proposed by Chorley and Kennedy [23] with their approach to classify systems on the basis of their structural complexity. Although the two authors do not define complexity, they seemingly understood complexity as a high number of elements and a high degree of relations between them (cf. [67]). The classification by Chorley and Kennedy [23] has now been valid for more than 30 years, and modifications referred to adding time dependency of processes as classification criteria, so that the classification was extended by the two types of macro- and mesoscale morphogenetic systems (cf. [68, 69]). The original classification comprises four types of systems (for a detailed description please refer to [23, 70, p. 78]).

1. The morphological system: This type of system is solely composed of the physical features [e.g. geometry (slope angle etc.) and composition (grain size distribution etc.)] which form a structural network. One important characteristic of morphological systems are the positive and negative feedback loops that determine how (and if) changes of single variables have an influence on the morphology.
2. The cascading system: Cascading systems consist of a number of coupled subsystems. The coupling may take place through different geomorphic processes (e.g. avalanches, debris flows) that transfer mass and/or energy to another subsystem. The output of one system thus is the input to another system. If and how long mass and energy can be partially or completely stored depends on regulators: Either, mass and energy are partially or completely stored, and/or they are remobilized after differing residence times. (Re-)Mobilized mass and/or energy then passes through the systems as throughput and finally leave the system as output. Cascading systems thus focus on quality and quantity of input/output-relations.
3. The process-response system: Within process-response systems, morphological and cascading systems are interlinked via regulators and storages. Here, morphological features either serve as decisive regulators, or variables of the

cascading system and variables of the morphological system are strongly correlated. The interfingering of the two system types is often realized by negative feedbacks. Due to these mechanisms process-response systems have the property of self-regulation so that they can adapt to changing framework conditions.
4. The control system: Regulators and other key variables of process-response systems allow human interventions to change the functioning as well as the equilibrium states of single morphological variables within systems. If these interventions are conscious and controlled measures, the social decision system and the process-response system form a control system. At least in theory it is assumed that these interventions or the human control is proceeding to more and more informed measures so that unforeseen and unwanted consequences are constantly being reduced.

To conclude it can be stated that within the sketched technical understanding that is typical for geomorphology, a system performs a transformation process with inputs and outputs (the only exception are morphological systems). This again fosters an arbitrarity of the systems approach, since in the end, any interrelation can be analysed according to its in- and output-relations. Hence, in- and outputs are by no means a characteristic feature that only applies to systems (also cf. [58]). From this perspective most geomorphic systems are

1. open, as energy and mass is transferred. This transfer of energy and mass produces a hierarchic structure of systems (cf. [23]), which
2. are coupled or, respectively, interfingered as the output of one system can be the input to another system, and
3. structured, since the in- and outputs are capable of influencing and changing system components so that the structure can not only be determined by internal interactions but also by interactions with the environment [71, p. 23] (cf. Fig. 3.2). Figure 3.2 shows that the material flows are focussed and only to a lesser extent the structures of the system. Furthermore, often a certain hierarchy of flows is regarded as structure. Hence, systems can neither be unequivocally delimited nor is there an unequivocal structure of the system. Both always depend on the specific research question.

These three aspects have essential implications for the system approach as they, first, found the perception of a complex interrelationship of geomorphic systems, and as they, second, illustrate why geomorphic systems are perceived as environmentally determined or at least determinable: It is a sheer metabolistic approach that quasi automatically results in an understanding according to which system behaviour depends on the functioning of the exchange between system and environment.

The above classification reaches its limits if e.g. Christopherson's [33] definition of systems is taken as a basis. It focuses on mass flows that are by definition not taking place in morphological systems—according to Christopherson these thus are not systems. Classification and definition are incompatible. If other, more

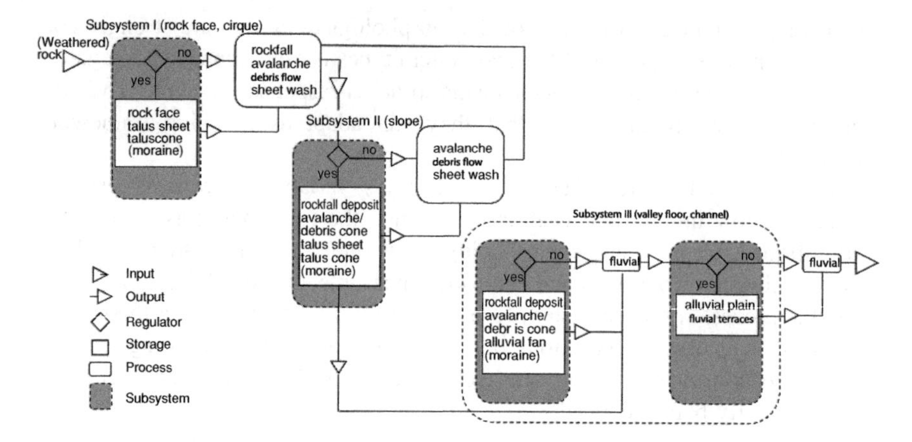

Fig. 3.2 An example for the geomorphological perspective on systems (here: cascading system). The focus is on input–output-relations and simultaneous consideration of structures and processes ([65], modified)

recent scientific definitions of systems and distinguishing features are used, respectively, the fracture or incompatibility of the different system approaches becomes even more obvious. This becomes particularly clear in comparison with the above-mentioned autopoiesis concept of the biologists Maturana & Varela [54] (also cf. [50]). Chorley and Kennedy [23] grant the environment a certain determining influence, and due to the close interlinkage and coupling of different systems and/or their energy and mass fluxes, respectively, it is difficult to distinguish environment and system in the end (see above). This coupling and interfingering does not only take place on the level of structure, but also on the process level as no such thing as a process exists (or mode of operation) that is characteristic for one specific type of system.

Within the traditional system understanding geomorphic systems are neither self-referential nor self-organizing, at best they are self-regulating. From this perspective the above classification can be interpreted as an attempt to transfer the perceived 'interrelation of everything' to a manageable order. This organising attempt becomes obsolete as soon as a second order system approach is applied since it offers a definite distinguishing feature for the definition and distinction of systems. As it creates itself the system reduces its own complexity through autopoiesis and the associated creation of order (structures)—the system is always less complex that its environment. But how does an autopoietic system maintain relations to its environment? Obviously, due to the self-created system boundary these environmental relations have to be organized in a different way than in principally open systems such as geomorphic systems. Against this background, the main questions are how the respective system organises its environmental relations, and, most of all, how strong is the influence of these environmental relations on the system behaviour.

References

1. Dikau R (2005) Geomorphologische Perspektiven integrativer Forschungsansätze in Physischer Geographie und Humangeographie. In: Wardenga U, Müller-Mahn D (eds) Möglichkeiten und Grenzen integrativer Forschungsansätze in Physischer Geographie und Humangeographie. Forum ifl. Leibniz-Institut für Länderkunde, Leipzig, pp 91–108
2. Cox NJ (2007) Kinds and problems of geomorphological explanation. Geomorphology 88(1–2):46–56
3. Bak P (1996) How nature works. The science of self-organised criticality. Copernicus Press, New York, p 212
4. Hergarten S (2003) Landslides, sandpiles, and self-orgnaized criticality. Nat Hazard Earth Syst Sci 3(3):505–514
5. Keiler M (2011) Geomorphology and complexity—inseparably connected? Zeitschrift für Geomorphologie 55(3):233–257
6. Phillips JD (1992) The end of equilibrium? Geomorphology 5(3–5):195–201
7. Phillips JD (1992) Nonlinear dynamical systems in geomorphology revolution or evolution? Geomorphology 5:219–229
8. Phillips JD (2003) Sources of nonlinearity and complexity in geomorphic systems. Prog Phys Geogr 27(1):1–23
9. Phillips JD (2006) Deterministic chaos and historical geomorphology: a review and look forward. Geomorphology 76:109–121
10. Phillips JD (2006) Evolutionary geomorphology: thresholds and nonlinearity in landform response to environmental change. Hydrol Earth Syst Sci 10:731–742
11. Schumm SA (1991) To interpret the earth. Ten ways to be wrong. Cambridg University Press, Cambridge
12. Kuhn TS (1962) The structure of scientific revolutions. University of Chicago Press, Chicago
13. Sternberg RJ, Conway BE, Ketron JL, Bernstein M (1981) People's conceptions of intelligence. J Pers Soc Psychol 41(1):37–55
14. Furnham A (1988) Lay theories. Everyday understanding of problems in social sciences. Pergamon Press, New York
15. Eibl K (1976) Kritisch-rationale Literaturwissenschaft. Grundlagen zur erklärenden Literaturgeschichte. München
16. Egner H, von Elverfeldt K (2009) A bridge over troubled waters? Systems theory and dialogue in geography. Area 41(3):319–328
17. Scheidegger AE (1992) Limitations of the system approach in geomorphology. Geomorphology 5:213–217
18. von Elverfeldt K, Keiler M (2008) Offene Systeme und ihre Umwelt—Systemperspektiven in der Geomorphologie. In: Egner H, Ratter BMW, Dikau R (eds) Umwelt als System—System als Umwelt? Systemtheorien auf dem Prüfstand. Oekom, München, pp 75–102
19. von Elverfeldt K, Glade T (2011) Systems theory in geomorphology. A challenge. Zeitschrift für Geomorphologie 55(3):87–108
20. Seiffert H, Radnitzky G (ed) (1994) Handlexikon zur Wissenschaftstheorie. dtv Wissenschaft, München, p 502
21. Mayhew S, Penny A (eds) (1992) The concise Oxford Dictionary of Geography. Comprehensive coverage in one volume of both human and physical geography. Oxford Reference, Oxford
22. Brunotte E, Gebhardt H, Meurer M, Meusburger P, Nipper J (eds) (2002) Lexikon der Geographie. In vier Bänden, Heidelberg, Berlin
23. Chorley RJ, Kennedy BA (1971) Physical geography—a systems approach. London
24. Howard AD (1965) Geomorphological systems–equilibrium and dynamics. Am J Sci 263(4):302–312
25. Bull WB (1991) Geomorphic responses to climatic change. Oxford University Press, Oxford, p 326

26. Huggett RJ (2003) Fundamentals of geomorphology. Routledge, London, p 336
27. Bailey KD (2007) Boundary maintenance in living systems theory and social entropy theory. In: The 51st annual meeting of the international society for the systems sciences. Integrated Systems Sciences: Systems Thinking, Modeling and Practice, Tokyo, pp 1–15
28. Baker VR, Pyne S (1978) G. K. Gilbert and modern geomorphology. Am J Sci 278:97–123
29. Honig JM (1999) Thermodynamics
30. White ID, Mottershead DN, Harrison SJ (1992) Environmental systems. Stanley Thornes (Publishers) Ltd, Cheltenham, p 616
31. Kennedy BA (1979) A naughty world. Trans Inst Br Geog 4(4):550–558
32. Werner BT, McNamara DE (2007) Dynamics of coupled human-landscape systems. Geomorphology 91:393–407
33. Christopherson RW (2006) Geosystems: An introduction to physical geography, p 689
34. von Foerster H (2006) Sicht und Einsicht. Versuche zu einer operativen Erkenntnistheorie. Carl-Auer, Heidelberg, p 233
35. Moran MJ, Shapiro HN (1992) Fundamentals of engineering thermodynamics. Wiley, New York
36. Thorn CE, Welford MR (1994) The equilibrium concept in geomorphology. Ann Assoc Am Geogr 84(4):666–696
37. Sattelmayer T (2009) Thermodynamik I: Grundkonzepte und Definitionen. Available at http://www.td.mw.tum.de/tum-td/de/lehre/thermo_1/download/D-folien/B-Handout-Kapitel2.pdf. Accessed on 27 Nov 2009
38. Fuchs P (1992) Niklas Luhmann–beobachtet. Eine Einführung in die Systemtheorie. Westdeutscher Verlag, Opladen, p 219
39. Dubrovsky V (2004) Toward system principles: general system theory and the alternative approach. Syst Res Behav Sci 21:109–122
40. Larses O, Elkhoury J (2005) Views on general systems theory. TRITA-MMK: 2005:10, Mechatronics Lab, Department of Machine Design, Royal Institute of Technology, KTH, Stockholm
41. Prigogine I, Stengers I (1990) Entwicklung und Irreversibilität. In: Niedersen U, Pohlmann L (ed) Selbstorganisation und Determination. Selbstorganisation. Jahrbuch für Komplexität in den Natur-, Sozial- und Geisteswissenschaften. Duncker and Humblot, Berlin, pp 3–18
42. Spencer-Brown G (1996) Wahrscheinlichkeit und Wissenschaft. Carl Auer, Heidelberg, p 142
43. Chalmers AF (2001) Wege der Wissenschaft. Einführung in die Wissenschaftstheorie. Springer, Heidelberg, p 236
44. Schneider ED, Sagan D (2005) Into the cool. Energy flow, thermodynamics, and life. The University of Chicago Press, Chicago and London, p 362
45. Luhmann N (1986) Ökologische Kommunikation–Kann die moderne Gesellschaft sich auf ökologische Gefährdungen einstellen? Westdeutscher Verlag, Opladen, p 275
46. von Foerster H (1984) Observing systems. Intersystems Publications, Seaside, p 331
47. Spencer-Brown G (1997) Laws of form. Gesetze der Form. Bohmeier Verlag, Lübeck, p 200
48. Luhmann N (2008) Soziale Systeme. Grundriß einer allgemeinen Theorie. Suhrkamp, Frankfurt/Main, p 674
49. Luhmann N (2006) Einführung in die Systemtheorie. Heidelberg, p 347
50. Egner H (2008a) Gesellschaft, Mensch, Umwelt–beobachtet. Ein Beitrag zur Theorie der Geographie. Erdkundliches Wissen. Franz Steiner, Stuttgart, p 208
51. Luhmann N (1987) Soziale Systeme. Grundriß einer allgemeinen Theorie. Suhrkamp, Frankfurt/Main, p 674
52. Luhmann N (1997a) Selbstreferentielle Systeme. In: Simon FB (ed) Lebende Systeme. Wirklichkeitskonstruktionen in der systemischen Therapie. Suhrkamp, Frankfurt/Main
53. Maturana HR (1982) Erkennen: Die Organisation und Verkörperung von Wirklichkeit. Ausgewählte Arbeiten zur biologischen Epistemologie. Wissenschaftstheorie, Wissenschaft und Philosophie. Vieweg, Braunschweig/Wiesbaden, p 322

54. Maturana HR, Varela FJ (1984) Der Baum der Erkenntnis. Die biologischen Wurzeln des menschlichen Erkennens, p 280
55. Maturana HR, Varela FJ (1982) Autopoietische Systeme: Eine Bestimmung der lebendigen Organisation. In: Maturana HR (ed) Erkennen: Die Organisation und Verkörperung von Wirklichkeit. Ausgewählte Arbeiten zur biologischen Epistemologie. Wissenschaftstheorie, Wissenschaft und Philosophie. Vieweg, Braunschweig/Wiesbaden, pp 170–235
56. Bühl WL (1987) Grenzen der Autopoiesis. Kölner Z für Soziologie und Sozialpsychologie 39:225–254
57. Roth G (1986) Selbstorganisation—Selbsterhaltung—Selbstreferentialität. In: Dress A, Henrichs H, Küppers G (ed) Selbstorganisation. Die Entstehung von Ordnung in Natur und Gesellschaft. Piper, München, Zürich, p 149–180
58. Allefeld C (1999) Erkenntnistheoretische Konsequenzen der Systemtheorie. Die Theorie selbstreferentieller Systeme und der Konstruktivismus. Master Thesis, Freie Universität Berlin, Berlin, p 85
59. Eigen M (1971) Selforganization of matter and the evolution of biological macromolecules. Die Naturwiss 58(10): 465–523
60. Maturana HR (1994) Was ist Erkennen? Piper, München, p 244
61. Lippe zur R (1994) Denken und Leben. Essay zur Einführung von Rudolf zur Lippe. In: Maturana HR (ed) Was ist Erkennen? Piper, München, pp 7–23
62. Jantsch E (1979) Die Selbstorganisation des Universums. Vom Urknall zum menschlichen Geist. Hanser Verlag, Darmstadt, p 464
63. Maturana HR (1980) Autopoiesis: reproduction, heredity and evolution. In: Zeleny M (ed) Autopoiesis, dissipative structures and spontaneous social orders, Boulder, pp 45–79
64. Zeleny M (ed) (1980) Autopoiesis, dissipative structures and spontaneous social orders. In: AAAS selected symposium 55, Westview, Boulder, p 149
65. Egner H (2010) Theoretische Geographie. Wissenschaftliche Buchgesellschaft, Darmstadt, p 144
66. Pitty AF (1971) Introduction to geomorphology. Methuen and Co, Norwich, p 526
67. Pigliucci M (2000) Chaos and complexity. Should we be sceptical? Sceptic 8(3):62–70
68. Dikau R (1998) The need for field evidence in modelling landform evolution. In: Hergarten S, Neugebauer HJ (ed) Process modelling and landform evolution. Lecture notes in earth sciences, Springer-Verlag, Heidelberg, pp 3–12
69. Slaymaker O (1991) Mountain geomorphology: a theoretical framework for measurement programmes. In: Crozier MJ (ed) Geomorphology in unstable regions. Catena, Cremlingen, pp 427–437
70. von Elverfeldt K, Keiler M (2008) Offene Systeme und ihre Umwelt–Systemperspektiven in der Geomorphologie. In: Egner H, Ratter BMW, Dikau R (eds) Umwelt als System–System als Umwelt? Systemtheorien auf dem Prüfstand. Oekom, München, pp 75–102
71. von Bertalanffy L (1950) The theory of open systems in physics and biology. Science 111(2872):23–29
72. Schrott L, Niederheide A, Hankammer M, Hufschmidt G, Dikau R (2002) Sediment storage in a mountain catchment: geomorphic coupling and temporal variability (Reintal, Bavarian Alps, Germany). Z Geomorphol 127:175–196

Chapter 4
Second Problem Area: Openness and Determinacy

The greatest difficulties lie where we are not looking for them.
Goethe

4.1 The 'First Law of Geography'

As described above the 'unlimited' geomorphological system theoretical perspective only has a minor ordering potential—everything can be addressed as a system, and furthermore, everything that is addressed has open boundaries. From this point of view the world thus appears as a continuum, and any delimitation of 'meaningful' unities is arbitrarily brought forth by an observer ([1, p. 1], also cf. [2, p. 17]). Hence, geomorphology tends to perceive a world in which "everything is connected with everything else" ("1st law of geography", [3], personal communication). Everything 'is system'. Therefore, any system element and any 'element of the environment' can again be viewed as a system of its own, resulting in a system theory that can be recursively applied to every aggregation level (scale) (cf. [4, p. 11]). Depending on which system elements and their respective relations are supposed to be studied not only do the system boundaries vary, but also the interactions with its environment. As a result, within classical geomorphological system thinking the system is perceived as being embedded in an environment (also cf. [5]) and, respectively, everything is perceived as being connected to everything else.

In this context, Hard [9] states that system thinking is nothing more than old wine in new bottles as the traditional landscape concepts are simply transformed to concepts of systems. For example, this surfaces when Renwick [10] uses a definition of landscape that is equivalent to the geomorphological definition of the term 'system'. According to Hard [9, p. 94], system theory is suggestive of a very extensive complexity and, respectively, of an interrelation of all phenomena, which however, does not lead any further than to the conclusion that, in the end,

---

Some of the thoughts presented here have already been published: Egner and Elverfeldt [6], Elverfeldt and Keiler [7], as well as Elverfeldt and Glade [8].

K. von Elverfeldt, *System Theory in Geomorphology*,
Springer Theses, DOI: 10.1007/978-94-007-2822-6_4,

within a geo-complex, geosystem or ecotope everything is connected to everything else in an alternating manner. Hard claims that these thoughts are only scientifically useful to a certain degree.

Thus, it is possible to arbitrarily define where a specific system ends and where the environment begins (cf. Chap. 3). An unequivocal delimitation of the system by the system itself is only possible if the system *is closed in itself* [11, pp. 213–214]. However, the classification of Chorley and Kennedy [1] shows that geomorphic systems are necessarily open for the exchange of mass and energy and are, furthermore, even defined through these transformation processes. In order to preserve itself an open system exchanges mass and energy with its environment.[1] Thus, the system approach can be applied to the classical geomorphological research objects such as catchments, slopes, river sections etc. However, the assumption that system boundaries are generally (temporal and spatial) open is not correct: Simplified, a landslide-dammed river could for example be regarded as temporally closed for outputs, but open for inputs. Within geomorphology, this is generally described by the concepts of coupling and decoupling, respectively, of systems or parts of systems.

4.2 Environmental Determinacy of Geomorphic Systems

Seen from the viewpoint of geomorphology, the environment is capable of determining the behaviour of the system, which is a consequence of the basic openness of geomorphic systems. After all, systems and their organisation are not only influenced (temporarily or continuously) by inputs whenever the input exceeds a certain threshold [14] and upon which the system then shows some reaction (e.g. when a rainfall event of a certain magnitude triggers a landslide). Rather, mass and energy inputs continuously influence the system's structure (cf. [15]), and therefore also its inner disposition (intrinsic thresholds, cf. [14]) to react to disturbances at all[2]. This can also be interpreted as 'memory' or history of the system. At the same time, as soon as it reaches the system, the imported mass and energy is also influenced and transformed by the system structure and the

[1] In this context it might be interesting to refer (even if as a side aspect) to Einstein's formula ($E = mc^2$) from the year 1905. It postulates the equivalence of mass and energy: mass (m) can thus be regarded as a concentrated form of energy (E), multiplied by the square of the speed of light (c) (also cf. [12, 13]. As a consequence, the difference between open (transfer of mass and of energy) and closed (transfer of energy) systems disappears—in sensu stricto systems can only be either isolated or open.

[2] In his article on geomorphic thresholds Schumm [14] distinguishes extrinsic and intrinsic thresholds. Extrinsic thresholds are those thresholds within a system, which are exceeded because of the change of an external variable (e.g. vegetation cover). In contrast, intrinsic thresholds are exceeded, although the framework conditions are not changing. This can be explained by the continuous throughput of mass and energy (and information) that constantly changes the structure of the system.

internal processes. For example, the system determines which input and how and when it leaves the system: The kind of system reaction influence the kind of output. Abstractly speaking, not only do mass and energy cascade from one system to the next, but the respective system reactions are also passed on as information (understood as a utilizable pattern of mass and energy). By engraving past events and system states, systems quasi have a memory, which determines their reactions as well as those of coupled systems. The transfer of energy, information, and mass can happen directly or with delay, depending on whether the systems are coupled by permanent, periodic, or episodic processes. This, in turn, depends on the temporal, but also spatial scale of observation (on the concept of scale cf. [7]).

The basic openness of geomorphic systems seems to be at the core of the problem of defining and delimiting systems. For example, the mathematician, protagonist of systems theory, and author of the second report of the Club of Rome, Mihailo Mesarovic (194:8), asked how a system can be different from another system when it is open. This is further complicated by the fact that the same processes can be found inside and outside of many geomorphic systems. Hence, this system thinking automatically leads to systems that are not only open for mass and energy, but also spatially (meaning that the system boundaries can be 'freely' shifted according to the research question) and processually variable.

Second order system theories open a completely different perspective on the openness und closeness of systems, including the degree to which the environment is capable of influencing the system (this has already been indicated in Sect. 3.1 on the definition and delimitation of systems). This perspective starts with the proposition that a system is a difference instead of a unity [16, p. 91]. Hence, the focus shifts towards a difference, the difference between system and environment. How this difference can be assessed has been answered with the concepts of autopoiesis and self-reference: Autopoietic systems create themselves and all of their elements and relate only to themselves with all their operations. Thus, the systems create boundaries with their own operations—and it is this different type of operation inside and outside a system that renders a system observable (cf. [16, p. 92]).

The German system theorist Niklas Luhmann (z. B. [16–18]) transferred this scientific concept of autopoiesis from living systems to two additional kinds of systems, social systems and psychic systems, the astrophysicist Erich Jantsch furthermore to chemical and physical systems ([19], cf. [20, p. 134]).[3] The respective autopoiesis for social and psychic system is communication and

[3] Here, it should maybe be stated that Maturana and Varela never approved of the propagation of their autopoiesis concept to systems other than biological systems (cf. e.g. the prologue of Erich Jantsch [19], as well as Ludewig and Maturana [21]). Especially in the case of Jantsch this seemed to be justified insofar as he never named the specific kind of operation (that is, the precise autopoiesis) of the chemical and physical systems. Jantsch equalises autopoiesis with self-reference, which does not seem to be sufficient. For more critiques of the propagation of the concept of autopoiesis, in particular of Luhmann, please refer to Bühl [22], Krüger [23] and Mingers [24], as well as the amusing dispute between Jantsch and Maturana in Simon [20, p. 134].

consciousness. That is, wherever there is communication we are dealing with a social system, where one thought follows the other, there is consciousness and thus a psychic system, and where we discover cell division we know that we are dealing with a living system. The system brings forth its entire phenomenology[4] through its operations [25], all its internal operations depend on the preceding operations and are, at the same time, the basis for all its future operations. This aspect is summarized by the concept of the 'operative closeness' of systems: Utilizing the terminology of George Spencer-Brown one would say that the system is always operating on the inner side of the form, that is, in itself, and not on the outside of the form. However, operating in the inside and not in the outside requires that an environment, an outside, exists. If expressed in simpler terms, this seems to be trivial, as it seems obvious that a system cannot operate within its environment, so that the operations happen always within the system. If system operations took place within the environment, the distinction of system and environment was sabotaged. This becomes less trivial, even surprising, if this thought is taken further in order to include the consequences: a system thus cannot utilize its own operations to get in contact with its environment. This is exactly the core of the concept of operative closeness. Right from the beginning, up to the end, or even seen as events, operations are only possible within the system. They cannot be used for linking to the environment, as then they need to become something else than system operations as soon as they cross the boundary [16, p. 92].

At this point, a further contradiction between first and second order system theory is revealed and which refers to the understanding of 'environment'. Whilst within geomorphology 'environment' is, in the end, everything that is not considered, within Luhmann's as well as Maturana's and Varela's perspective systems and environment are fundamentally different. After all, the environment neither has a specific *modus operandi* nor does it act self-referentially. Hence, environment can never be regarded as a system per se and in itself. In contrast, environment simply is the totality of external conditions that limit the forming and development of systems [26]. In this perspective, 'nature' or 'the earth' as a whole can also not be viewed as a system.[5] Environment is always seen in a very specific relation to system: System and environment either co-evolve—or not at all. Environment is system-relative, it is a negative-correlate of the system, environment simply is 'everything else' [17]. It thus cannot be referred to as *the* environment: Every system has its own and absolutely specific environment, thus there are many different environments. Thereby, other systems can also become part of the environment of a specific other system. But in contrast to the system, the environment is not an operational entity and does not have the ability to perceive,

[4] Maturana & Varela [25, p. 273] understand phenomenology as the totality of those phenomena, which are connected with the interactions of one class of entities.

[5] According to the logic of George Spencer-Brown, also the world as a whole cannot be a system as no distinction (to something else) could be applied (cf. [27, p. 87]). In other words: The environment of a system is everything except the system. Environment and system taken together always are the world [27, p. 91].

act on, or influence other systems. The environment only receives unity through the relation to the system: The environment does not have boundaries, just open horizons (cf. [17]).

If these explanations are taken seriously, neither system nor environment can directly access one another on the level of operations (i.e. processes). Referring to this, Maturana and Varela [28, p. 202] explain that autopoietic systems do not have inputs and outputs. However, this statement explicitly refers to the level of operations or to the entity of the system: Entity can neither be imported to nor exported from the system (cf. [29, p. 56]), whilst it is, as a matter of course, energetically open (also cf. [22]). Instead of transforming inputs to outputs the system transforms itself into itself, leading to a cycle [24], and this is exactly why the environment cannot determine system behaviour. At the same time, systems are structurally oriented towards their environment and cannot exist without it–in this sense, boundary preservation is system preservation [17, p. 35]. The question of traditional systems theory, whether we are dealing with closed or open systems, becomes within second order systems theory the question: How can self-referential, operatively closed systems create openness, that is take in information and energy from the environment (cf. [17, p. 25])? This is answered by the autonomy of autopoietic systems, which should not be confused with autarky (also cf. Sect. 3.1), or, to put it differently, that systems are merely *operatively closed* and not *energetically closed*. They rely on the throughput of energy—however, specific framework conditions are solely necessary conditions, not sufficient conditions for the system to exist at all. In a way, inputs and outputs are externalised, they are seen as prerequisite for the system's existence; its existence requires a physically functioning environment. Without sun, water and food there is no life on earth, and without humans there is no communication. The closeness of the systems due to autopoiesis thus should not be confused with self-sufficiency.

Thus, although a system needs a functioning environment the environmental conditions are not determining the system in the sense that they directly and causally determine or produce the structure of the system. The system is sensitive to the environment, but by no means reactive to it [22, 26]. A primacy of the environment over the system is not allowed for as the system does not have to adapt to its environment [24]. As soon as the system exists it is automatically adapted, and the way *how* exactly it adapts itself to the environment is determined by the system structure.[6] That is, autopoietic systems as a tree are open for energy, but these flows cannot change cell division to another mode of operation. This also corresponds to our personal experience as living system: We need food (energy), but this food can only influence our functioning if it collides with our structures, for example when the food is spoiled. That this is no inevitable mechanism that solely depends on the input (i.e. the food) becomes obvious any time when of two

[6] Here, Jantsch [19, p. 75] goes even further by ascribing autopoetic systems a kind of primitive consciousness. As my understanding of consciousness fundamentally differs from this (I deny that self-reference is the only criteria for consciousness). I do not want to follow this thought.

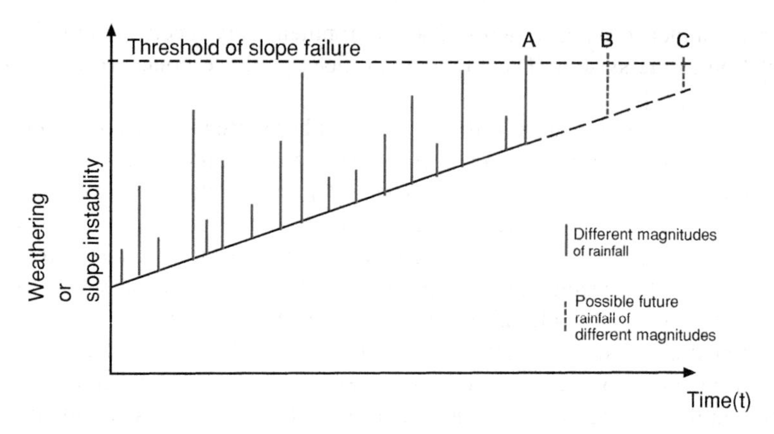

Fig. 4.1 A geomorphological example for the information selection of systems. A slope may not react to a heavy rainfall event whilst it does (a little) later to a relatively small event ([6, 14, 31], modified)

people who ate the same spoiled food, one gets more sick than the other (or maybe in contrast, not sick at all). Another example is medication: Time after time there are cases where medication has no effects or even contrary effects to the intended ones, e.g. in the case of medication allergies or strong adverse reactions that only occur with some people but by no means with all. These examples have in common that the input (the food or the medication) is the same, but that different systems react completely differently according to their internal structures. That is, the system defines according to its inner structure how and if at all it reacts to this irritation ('trigger'): It distinguishes what has to be taken as *order from noise* (cf. [30]). In this sense, irritation is always self-irritation as the system itself sets which environmental information is relevant and which it thus filters (selects) from the general and continuous noise. Irritations thus are internal constructions that result from a confrontation of the events with the system structures. Irritations are therefore not taking place within the environment but within the system itself.

This perspective can be analogically applied to geomorphic systems, the 'nutrition' of which is energy (and mass) from the environment. For example, within natural hazard research it is a well-known phenomenon that a slope does not react to a heavy rainfall event, but some years (days, weeks, months) later it does react to a much weaker event (cf. Fig. 4.1).

Mostly this is explained by the reduction of an intrinsic threshold, e.g. by continuous weathering, thus leading to a higher landslide disposition. A further example is an experiment at the University of Karlsruhe, where a dyke in an experiment channel should be brought to failure (cf. [31]). For this, the channel was slowly but constantly flooded. According to the underlying assumptions on the stability of slopes and embankments the dyke was expected to fail at a certain point—whereas it stayed stable. First order system theories do not offer plausible explanations for phenomena like this, as according to this theoretical approach the system is expected to react in accordance with the framework conditions. After all, the functioning of systems is completely (or at least critically) determined by the environmental conditions. In the light of the above thoughts these two examples could be understood differently: It depends on the inner structure, not on environmental conditions, how the slope or the dyke reacts. The degree of influence of the environment on the system therefore depends on the kind of utilized system approach.

How does a system as viewed from the perspective of second order system theory relate to its environment in order to continue its autopoiesis? In other words: How does structural change become possible without a loss of the mode of organisation?[7] According to Maturana and Varela [25, p. 84] this change continuously takes place within the system, be it due to interactions with the environment or due to internal dynamics. A specific characteristic of autopoietic systems thus is change, and not standstill, for instance (as it is expressed within equilibrium concepts, cf. Chap. 6), since the continuous interactions with the environment are always in accordance with the system structure, which, in turn, in the course of its internal dynamics is again subject to continuous changes. The universal result from this is that the ontogenetical change never stops.

From the system's standpoint it is completely irrelevant whether these interactions are taking place with another system or with a non-autopoietic environment. What is relevant, however, is that the environment only has the capability of irritating the system, but not of determining or instructing its actions. In the reverse, this is also true for the influence of the system on the environment. This results in a history of mutual and reciprocal structural changes [25], termed "structural coupling". The essential characteristics of this kind of coupling are the recursiveness, the stability and that these couplings take place on the level of the structures. Thus, structural coupling does not influence the type of self-reproduction and in consequence does not contradict the operative closeness due to autopoiesis. In more simplified terms, structural coupling can be understood as well-attuned ways of handling self-selected irritations. Further specific features of structural coupling are its—at the same time—restricting and enabling character, as well as the absence of causalities (structural coupling represents simultaneity, as system and environment are either created simultaneously or not at all). Structural changes within a system are complementary to changes in its environment, because they depend on the history of interactions. Consequently, every system has to be individually studied.

Summarizing, autopoietic systems have five major characteristics, which at the same time mark a distinction, i.e. a distinction to first order system theories (also cf. [24]):

[7] Maturana & Varela [25, p. 273] call the history of the structural change ontogenesis.

1. Autopoietic systems are self-reproductive cycles whereby they are operationally closed (but energetically open). Inputs and outputs are indeed taking place, but the focus is on the mode of organisation.
2. As a consequence of their self-creation, autopoietic systems are highly autonomous, but not autarkic. In other words, they are environment-sensitive, but not environment-determined. As soon as the necessary framework conditions are given, the system is automatically adapted as long as it continues its autopoiesis.
3. There is a distinct difference between structure and organisation. Maturana [32] gives the example of a chair[8]: If one leg of a chair is shortened, it still remains a chair—its functioning, its inner organisation remains, although its structure has been changed. However, if the chair is sawed through in the middle, its organisation is also changed: It is no longer recognizable as a chair (at best as a former chair) and cannot fulfil its function. Thus, organisation refers to the relations that have to be given so that a 'something' is exactly *this* something. By utilizing organisation we automatically and quasi intuitively define classes—e.g. the class 'chair', which is given by a specific relation of back, seat, legs, etc. (cf. [25]). In contrast to its mode of organisation, the structure of an autopoietic system can change with time. Mingers [24] refers to this as structural openness (with simultaneous operative closeness).
4. Any changes can only take place within the limits of the autopoiesis (otherwise, the system ceases its existence). Which changes are taking place is neither *solely* determined by the structure nor solely by the environment, but by a complex network of coincidence, structure, history, and environmental conditions. The term 'structural determinacy' is supposed to express that the system can only react within the limits of its own current structure to irritations from the environment—determinacy is thus understood in a broader sense than in the classical 'solely determined by something' (also cf. [22]).
5. Systems can be structurally coupled to other systems, whilst this is a process of mutual specification and not an adaptation of one system to another.

So far it has been shown that there exist at least two scientific system theories, which fundamentally differ in their basic assumptions. Within the perspective of geomorphological first order system theory, a basic assumption is that everything can be regarded as system, and the system boundaries can be defined according to the respective research question. Accordingly, system definitions generally refer to a given interdependence, i.e. an interaction of elements. That is, causalities are at the centre of attention, and as soon as such a causality cannot be established, this is blamed on the absence of process knowledge or other factors, but not on the (at least equally) potential lack of any causalities. A second central assumption is the openness of systems, whereby mass and energy flows are focused and thus turn to the central element of system analysis. Inputs are thereby capable of determining system behaviour: In this perspective, geomorphic systems are externally organised (environmental-determined) rather than self-organising, even though the way in

[8] This is a purely didactical example of Maturana—a chair does not exhibit autopoiesis!

which throughputs pass the system is seen as being influenced by the respective system state as well as the system configuration. Due to this close interfingering or coupling of systems to their environment, and due to the absence of definite system boundaries, the systems are perceived as being embedded within their environment.

The biological theory of autopoietic systems that has been presented as an example for a second order system theory within the natural sciences, however, is starting from completely different assumptions. Although systems are also seen as being energetically (and therefore also materially) open, this is not a central aspect to the system analysis: An existing metabolism simply is the prerequisite for the existence of a system. As we will see in the next chapter, only open systems are capable of building order in form of (symmetry-breaking) structures. Autopoiesis will be suggested as system- and thus structure-building process, or rather complex of processes. At the same time, autopoiesis is a sufficient definition criterion for the distinction of system and environment. Though systems thus remain theoretical constructions, this perspective offers a coherent rule for the definition of systems, as the rule stays the same regardless of the research question. With the introduction of a research-independent system boundary the influence of the environment is dramatically reduced. After all, any irritation has to manage to cross this boundary first, which is further, in a sense, 'buffered' by the fact that environment and system do not share the same mode of operation. Hence, systems within second order system theory are self-organising, operationally closed, and exhibit non-linear behaviour.

Still, geomorphology is not simply a sub-discipline of biology, and against this background many of these arguments could merely be relatable to the differences of disciplines and thus be rejected. However, the price to be paid would be the inherent incoherence of geomorphological systems theory, as this approach is mainly characterized by logical flaws as detailed earlier:

- On the one hand, geomorphology assumes that a system is an entity (of interdependencies between elements), on the other hand this entity cannot be specified. As delimitation 'method' solely 'common sense' is suggested.
- The definition of systems circularly refers to itself: Relations between elements render a system, but in order to delimit a system by this criterion, the elements as well as their relations already have to be known. In order to define something independently of itself a distinctive feature is needed, which, however, is not given here.
- As a consequence there exists no coherent principle for the delimitation of systems, thus, the system analysis becomes subjective and arbitrary to a certain degree.
- In this sense, the concept of systems applies to simply everything, and thus cannot be wrong. There are no specific and verifiable predictions or observations that can be excluded from the system approach.
- Geomorphological definitions of systems are partly contradicted by the common classification scheme, e.g. when on the one hand a system is defined by energy and mass flows, but on the other hand the classification allows for systems that

are defined by the mechanical relation of components (such as the dependency of the maximum slope angle of a dune on grain sizes).

- A classification principle that is based on 'the degree of complexity' only has a low organising potential and rather resembles an instruction for reductionism.
- Summa summarum, geomorphological systems theory only resembles a terminological transformation of old typological concepts, whereas a transformation of content is rather only indicated.

These limitations were made visible by utilizing the distinction of first order system theory vs. second order system theory. Previously, they have been blind spots of geomorphology. I (and maybe also the reader) now got curious: What are the consequences of these limitations? Are there further inconsistencies? Therefore, I thought it necessary to have a closer look at the basics of geomorphological system theory. This also seems to be reasonable as it has not become clear so far whether second order system theories offer real alternatives for geomorphology. After all, the foundation of geomorphology is physics, not biology. However, it is exactly at this point that a unifying aspect surfaces: Both disciplines cannot be reduced to physics, but at the same time they cannot contradict physical laws (amongst many, cf. [33–38]). The physical theory that is basic to all analyses of energy flows as well as of the evolution of systems is thermodynamics. Hence, it is thermodynamics with which geomorphological system theoretical concepts (as well as those of other sciences) have to be consistent. This is why the main features of thermodynamics will be explained in the next chapter.

References

1. Chorley RJ, Kennedy BA (1971) Physical geography—a systems approach. London
2. Dikau R (1996) Geomorphologische Reliefklassifikation und -analyse. Heidelberger Geographische Arbeiten, p 104
3. Phillips JD (2009) Changes, perturbations, and responses in geomorphic systems. Prog Phys Geogr 33(1):17–30
4. Larses O, Elkhoury J (2005) Views on general systems theory. TRITA-MMK: 2005: 10, Mechatronics Lab, Department of Machine Design, Royal Institute of Technology, KTH, Stockholm
5. Slaymaker O, Spencer T (1998) Physical geography and global environmental change. Adison Wesley Longman, Harlow, p 292
6. Egner H, von Elverfeldt K (2009) A bridge over troubled waters? Systems theory and dialogue in geography. Area 41(3):319–328
7. von Elverfeldt K, Keiler M (2008) Offene Systeme und ihre Umwelt—Systemperspektiven in der Geomorphologie. In: Egner H, Ratter BMW, Dikau R (ed) Umwelt als System—System als Umwelt? Systemtheorien auf dem Prüfstand. Oekom, München, pp 75–102
8. von Elverfeldt K, Glade T (2011) Systems theory in Geomorphology. A challenge. Zeitschrift für Geomorphologie 55(3):87–108
9. Hard G (1973) Zur Methodologie und Zukunft der physischen Geographien an Hochschule und Schule. Möglichkeiten physisch-geographischer Forschungsperspektiven. Geog Z (61): 5–35 (hier aus: Hard G (2003) Dimensionen geographischen Denkens. Aufsätze zur Theorie der Geographie, vol 2. Osnabrück, pp S87–111)

10. Renwick WH (1992) Equilibrium, disequilibrium, and nonequilibrium landforms in the landscape. Geomorphology 5:265–276
11. Scheidegger AE (1992) Limitations of the system approach in geomorphology. Geomorphology 5:213–217
12. Bailey KD (1994) Sociology and the new systems theory. Toward a theoretical synthesis. SUNY Press, Albany
13. Baumgärtner S (2005) Thermodynamic models. In: Proops JLR, Safonov P (eds) Modelling in ecological economics. Edward Elgar, Cheltenham
14. Schumm SA (1979) Geomorphic thresholds. Concept Appl Trans Inst Br Geog 4(4):485–515
15. Strahler AN (1952) Dynamic basis of geomorphology. Bull Geol Soc Am 63:923–938
16. Luhmann N (2006) Einführung in die Systemtheorie. Heidelberg, p 347
17. Luhmann N (1987) Soziale Systeme. Grundriß einer allgemeinen Theorie. Suhrkamp, Frankfurt/Main, p 674
18. Luhmann N (1998) Die Gesellschaft der Gesellschaft. Suhrkamp, Frankfurt/Main, p 1164
19. Jantsch E (1979) Die Selbstorganisation des Universums. Vom Urknall zum menschlichen Geist. Hanser Verlag, Darmstadt, p 464
20. Simon FB (1997) Kreuzverhör: Fragen an Heinz von Foerster, Niklas Luhmann und Francisco Varela. In: Simon FB (ed) Lebende Systeme: Wirklichkeitskonstruktionen in der systemischen Therapie. Suhrkamp, Frankfurt/Main, pp 131–147
21. Ludewig K, Maturana HR (2006) Gespräche mit Humberto Maturana. Fragen zur Biologie, Psychotherapie und den „Baum der Erkenntnis" oder: Die Fragen, die ich ihm immer stellen wollte. http://www.systemagazin.de/bibliothek/texte/ludewigmaturana.pdf. Last accessed 15 Jan 2012
22. Bühl WL (1987) Grenzen der Autopoiesis. Kölner Zeitschrift für Soziologie und Sozialpsychologie 39:225–254
23. Krüger H-P (1990) Luhmanns autopoietische Wende. Eine kommunikationsorientierte Grenzbestimmung. In: Niedersen U, Pohlmann L (eds) Selbstorganisation und Determination. Selbstorganisation. Jahrbuch für Komplexität in den Natur-, Sozial- und Geisteswissenschaften. Duncker and Humblot, Berlin, pp 129–148
24. Mingers John (2002) Can social systems be autopoietic? Assessing Luhmann's social theory. Soc Rev 50(2):278–299
25. Maturana HR, Varela FJ (1984) Der Baum der Erkenntnis. Die biologischen Wurzeln des menschlichen Erkennens. Bern, München, p 280
26. Luhmann N (1986) Ökologische Kommunikation—Kann die moderne Gesellschaft sich auf ökologische Gefährdungen einstellen? Westdeutscher Verlag, Opladen, p 275
27. Fuchs P (1992) Niklas Luhmann—beobachtet. Eine Einführung in die Systemtheorie. Westdeutscher Verlag, Opladen, p 219
28. Maturana HR, Varela FJ (1982) Autopoietische Systeme: eine Bestimmung der lebendigen Organisation. In: Maturana HR (ed) Erkennen: die Organisation und Verkörperung von Wirklichkeit. Ausgewählte Arbeiten zur biologischen Epistemologie. Wissenschaftstheorie, Wissenschaft und Philosophie. Vieweg, Braunschweig/Wiesbaden, pp 170–235
29. Luhmann N (1995) Soziologische Aufklärung. Die Soziologie und der Mensch, vol 6. Westdeutscher Verlag, Opladen, p 275
30. von Foerster H (1960) On self-organizing systems and their environments. In: Yovits MC, Cameron S (eds) Self-organizing systems. Pergamon Press, New York, pp 31–50
31. Bell R (2007) Lokale und regionale Gefahren- und Risikoanalyse gravitativer Massenbewegungen an der Schwäbischen Alb. PHD Thesis, University of Bonn, Bonn, p 270
32. Maturana HR (1994) Was ist Erkennen? Piper, München, p 244
33. Kauffmann S (2000) Investigations. Oxford University Press, Oxford u.a., p 287
34. Prigogine I (1985) Vom Sein zum Werden. Zeit und Komplexität in den Naturwissenschaften. Piper, München, p 304
35. Prigogine I, Stengers I (1981) Dialog mit der Natur. Neue Wege naturwissenschaftlichen Denkens. Piper, München, p 314

36. Prigogine I, Stengers I (1993) Das Paradox der Zeit. Zeit, Chaos und Quanten. Piper, München, p 338
37. Schneider ED, Sagan D (2005) Into the cool. Energy flow, thermodynamics, and life. The University of Chicago Press, Chicago & London, p 362
38. Schrödinger E (1944) What is life? The physical aspect of the living cell. Cambridge University Press, New York, p 32

Chapter 5
Third Problem Area: The Physical Basis

Even the highest towers begin from the ground.
Chinese proverb

5.1 Thermodynamics, not Mechanics!

In physics, three levels of consideration exist that cannot be reduced to each other (cf. [1, p. 55ff]): mechanics or dynamics, respectively, thermodynamics, and non-linear thermodynamics. Each of these levels focuses on a specific object, which results in findings of different scopes or significance. As will be shown in the following, mechanics and/or dynamics take place on the particle level and thus do not allow for statements on a system as a whole. Thermodynamics, in contrast, is a macroscopic approach, which, however, focuses static systems whilst the evolution of systems cannot be covered. How a system behaves, i.e. develops with time, can only be analysed by means of non-linear thermodynamics. Hence, it is equally important for geomorphology to be aware of the level of respective studies: For example, if the equilibrium of forces is analysed as in the case of slope stability analyses, this study is taking place on the level of mechanics, not allowing for any interpretations of a system state. Sediment budgets, on the other hand, are examples of investigations on the level of thermodynamics, as they focus flows of matter and therefore energy. Still, no conclusions on the evolution of systems can be made. Studies on self-organisation, as for river systems, have to focus on the whole system and infer how the system behaves and may behave in future and thus have to be based on non-linear thermodynamics.

In order to clearly distinguish these three levels of investigation and in order to know the level on which conclusions and observations have to take place, these three approaches are detailed in the following.

K. von Elverfeldt, *System Theory in Geomorphology*,
Springer Theses, DOI: 10.1007/978-94-007-2822-6_5,

5.1.1 First Level of Investigation: Mechanics

Together with dynamics, classical mechanics (also referred to as Newtonian mechanics) is the first level of investigation (cf. [2]). Within kinetics, which is a subdomain of mechanics, the movement of bodies is analysed. In order to do so, the positions and kinetics of a body as well as their changes with time have to be determined. The movement of the bodies is expressed by trajectories. The travelled path is reversible, so if the body travelled from A to B it can be lead back from B to A. Thereby, a directed arrow of time does not exist (also cf. [3]). The movement of a previously resting body can only be initiated by an external force. Such forces, or interactions between different bodies are studied within dynamics. For example, an equilibrium of forces or *mechanical equilibrium* can be visualised by scales or by a sphere on a plane (*indifferent equilibrium*), on a crest (*unstable equilibrium*), or in a depression (*stable equilibrium*) (cf. [4]). If the body is on an inclined surface and therefore not in equilibrium—meaning that the sum of all forces and torques is not zero—the acting forces will move it in the direction of a new equilibrium. As soon as the body reaches equilibrium, no changes will occur without external impact—no mechanical work can be performed. This mechanical equilibrium thus is the basis for several classical geomorphological questions such as slope stability analyses, questions of the transport capacity of rivers, or mass balances of glaciers. Examples for the incorrect extension of the scope of results that can be achieved by mechanical or dynamical analyses are the concepts of grade and of dynamic equilibrium (cf. Chap. 6), where mechanical analyses are used for the descriptions of systems as a whole. In some sense, a system that has been reduced to its interactions is reduced again to the forces, which are acting on single parts, whilst in reverse (at least) one aggregation level is leapfrogged, consequently leading to conclusions on contexts that have not even been studied.

5.1.2 Second Level of Investigation: Thermodynamics

If it is not the acting forces that are to be studied, for example if not the slope stability is to be focused, but instead a sediment budgeting, thermodynamics provide the basis for studies. In contrast to mechanics, thermodynamics operate on the macroscopic level, that is, the interplay of a variety of particles is studied. Processes, or rather energy flows, and statistical averages are focused. These averages are expressed by state variables such as temperature or pressure. State variables are characterised by the fact that they are only dependent on the (initial and end) state of a system, but they do not depend on the way in which they move from one state to the next (cf. [4, 5]). This is mathematically expressed by the ring integral over such a factor resulting zero.[1]

Due to the macroscopic point of view, in a thermodynamic system several equilibrium states are thinkable: If we transfer the above issue of an equilibrium of forces to an isolated system (with many parts), the mechanical equilibrium can be equalled to constant pressure within the system (cf. [6]). The *thermal equilibrium*, as expressed by the zeroth law of thermodynamics (cf. [4]), is reached by an isolated system after sufficient time: Isolated from any external influences, a system would not show any thermal gradients anymore (cf. [7]). As the temperature is a measure for the average kinetic energy of particles, thermal equilibrium means equal average velocity of the particles that move within a system. Isolated systems thus not only reach a mechanic equilibrium after a given amount of time, but also a thermal equilibrium.

So far, within this fictive isolated system, thermal and mechanical changes are impossible (that is, no thermal or mechanical work can be performed), however, chemical work can. The notion of *chemical equilibrium*, however, might be rather misleading as in the end the amount of particles is studied. Chemical equilibrium is reached as soon as two different groups of particles are completely mixed, so that the particle concentration is evenly distributed throughout the system, e.g. for the dissolution of salts or lime (for example, cf. [8]). A global, or absolute thermodynamic equilibrium is only reached when the system is in chemical, thermal and mechanical equilibrium cf. [6, 7].[2] The thermodynamic equilibrium that is characterized by maximum entropy (see below) thus can only be reached by isolated systems—and after sufficient time.

The question whether a system is in equilibrium or not is, in the end, the question whether a system is capable of performing work. For example, a system in thermal equilibrium cannot perform thermal work. Energy can occur in different forms, e.g. as kinetic, potential, or chemical energy. Energy can thereby be transferred from one form to the other with heat flow and work being the transfer processes. Though it should be noted that thermal energy is the kind of energy, which is the least utilizable for further work (cf. [10], pp. 26–27).

How thermodynamic systems develop was expressed for the first time by the second law of thermodynamics around the middle of the ninetieth century: Any

[1] State variables such as temperature are defined by the fact that their derivative (e.g. dT) is a complete derivative, or rather that the sum of all changes (e.g. of dT) on a path that is closed in itself (closed–loop integral) is zero. This again is independent of which path has been chosen. 'Path' here means the sequence of different system states (cf. [4, p. 49]).

[2] At this point it is important that the law of conservation of energy is not violated, or rather that it is only allowed in this way: the total energy within the system stays the same, but energy in the form of entropy cannot be utilized for work (cf. [5]). This 'loss of quality' of energy due to energy consumption can be understood more easily by means of an analogy to water consumption (cf. [9]): Used water, e.g. after laundry, has a lower quality than before. If work is done in a system, energy is consumed and the quality of energy is lowered. A measure for the quality of the energy is energy that thus is, at the same time, a measure for the distance from thermodynamic equilibrium (the less energy, the closer to equilibrium) (cf. [10, p. 32]).

thermodynamic system has a state variable, the entropy. It can be calculated with reversible processes. In an isolated system, real processes increase the entropy within the system [4, p. 56].[3,4]

As entropy, simplified, is a measure for that part of the total energy of a system, which is not capable of working, thermodynamics thus also deals with the quality of energy which is available for the system, as this influences the possible future states of the system. The entropy change during a thermodynamic process equals the amount of heat that has been taken in, divided by the system's temperature. In numerical terms, entropy equals the heat amount in Joule that is given to a reservoir at the temperature of 1°C. If the process is reversible, the same amount of heat per temperature is released so that the entropy change is zero in total. For irreversible processes, however, and this is true for all natural processes, the entropy increases. Thereby, energy is devalued; without further changes in the system it cannot be transferred to work (second law of thermodynamics). If we regard the development of a physical system—and this can be, for the start, just about anything: a river, a hurricane, or merely heated whale oil in a petri dish[5] –, then we *nolens volens* are also dealing with this second law of thermodynamics. In contrast to mechanics, with the second law of thermodynamics it becomes obvious that processes or developments are temporally directed as they are non-reversible due to energy consumption (production of entropy). On the one hand, with the increase of entropy the microscopic disorder of a system also increases, on the other hand the average state, which results from the uncontrollable microscopic behaviour, is completely determined and controlled by the framework conditions [11, p. 12]. The increased disorder is, at the same time, the more probable state than the ordered state, since for isolated systems it holds that states of less order are more probable than states of high order. This is of relevance insofar as, for example, the spontaneous heat transfer from cold to warm is not absolutely impossible, but 'only virtually impossible' [5, p. 610], also cf. [11, p. 7]. However, for macroscopic systems[6] the probability for such process is so low that it becomes impossible to distinguish between 'extremely improbable' and 'impossible' [5, p. 610]. Therefore, it can be concluded for the thermodynamic equilibrium and for entropy, that an isolated system without any external influences takes a state of high probability, low order and high entropy [5, p. 612]. The second law of thermodynamics thus states that all natural processes increase the entropy of the universe (also cf. [3])—provided that the universe is an isolated system. In contrast to earlier assumptions, however, it is currently assumed that the universe is indeed

[3] Central to the second law of thermodynamics is the distinction of reversible and irreversible ('real') processes. Only the latter contribute to entropy production as they devalue energy and thus produce the 'inferior' entropy that is incapable of working.

[4] There exists a multitude of equivalent formulations of the second law of thermodynamics.

[5] In his dissertation, this was the attempt of Henri Bénard, the name giver of the Bénard-cells.

[6] Macroscopic means that the system contains such a large numbers of molecules that microscopic fluctuations are negligible (cf. [7, p. 17]).

approaching thermodynamic equilibrium, i.e. 'heat death', but that it will not reach this state due to the expansion of the cosmos (cf. [12]).

Prigogine, the Nobel Prize Laureate for Chemistry of 1977, illustrates how far-reaching the consideration of entropy is for the understanding of systems, as it allows for distinguishing past and future: The second law of thermodynamics expresses the fact that irreversible processes introduce a direction of time. The positive direction of time is associated with an increase in entropy [13, p. 30].

Withal, the entropy function of isolated systems is a special kind of function as entropy can only increase with time: it is a so-called Lyapunov-function (e.g. [14]). The presence of a Lyapunov-function is a sufficient condition for the stability of a system—thus, the thermodynamic equilibrium is a stable system state. This can be explained by the fact that the correlations between the system elements equal zero, that is, fluctuations in one area cannot spread to other areas (cf. [11, p. 13]). For isolated systems—and only for these!—it can thus be stated that they are moving towards a state of thermodynamic equilibrium (it constitutes an attractor for isolated systems) that is characterized by an equalization of all gradients. In this situation, development only means that it leads to the demise of its own causes [3, p. 77].

Principally, the thought of Spencer-Brown, Foerster, Maturana and others can be continued: A system only emerges and persists if a difference to something else exists, the environment. In the case of the thermodynamic equilibrium, these differences no longer exist: Within the system there is no further differentiation in other (sub-) systems.

The basic problem of the application of thermodynamics to processes on the earth is that the laws of thermodynamics are principally valid only for isolated systems, that is, such systems that do not have any exchange with their environment. However, geomorphological (and all other) processes on earth do not meet this requirement as the sun supplies immense amounts of free energy for changes. Hence, the energy supplied by the sun drives all processes on earth, from weather conditions to erosion processes. At the same time, a massive disequilibrium exists between 'system earth' and 'system space' so that it is not surprising that equilibrium approaches fail, if (from this extremely macroscopic perspective) microscopic states such as the course of a river are viewed. Disequilibrium systems with a high excess of free energy tend to non-linearity (with according characteristics such as fractals and turbulences), which may explain the diversity of phenomena on earth. How we can deal with this basic problem will be explained in the following chapters on non-linear thermodynamics and on the theory of dissipative structures.

5.1.3 Third Level of Investigation: Non-Linear Thermodynamics

According to the above-mentioned, for all those systems that are not isolated, development has to be described and understood in other ways, as the thermodynamic equilibrium only is a very special type of system state [3, p. 77]. As soon as open systems are approached, one has to move from equilibria to non-equilibria. This is the third level of investigation in physics: non-linear thermodynamics. The problem of the laws of thermodynamics is, as already mentioned, that strictly speaking they are only valid for the linear state of dormancy, the thermodynamic equilibrium, where time is irrelevant (cf. [12]). Furthermore, thermodynamics is incapable of describing processes in time, as only initial and final states can be compared to each other [15]. Thus, an extension of thermodynamics seems necessary.

Based on works of the physicochemist and Nobel Prize Laureate Lars Onsager (1903–1976), Prigogine succeeded in transferring the formulations and basic assumptions of the second law of thermodynamics from isolated systems to open systems. In order to do so, open systems are for a start viewed close to equilibrium where linear relations between flows (irreversible processes such as heat flow, diffusion etc.) and forces (e.g. caused by temperature and/or potential differences), still can be assumed [13, p. 100]. Systems close to equilibrium are completely describable by their composition and framework conditions [11, p. 13]. For open systems, on the one hand the entropy production within the system and on the other hand the entropy transfer through the system borders is studied (cf. [7, 13]). In open systems, principally three different situations are possible: (1) increasing entropy with time, (2) constant entropy with time, and (3) decreasing entropy with time and order is established within the system. However, the second law of thermodynamics also states positive entropy production for open systems, however, this can be balanced or reduced by entropy export to the environment of the system.

The fact that the second law of thermodynamics is not violated by the transfer to open systems can be illustrated by viewing open systems as subsystems of an isolated system ('universe'). Then it becomes clear that, for example, an entropy reduction within a subsystem is associated with an entropy increase in the environment or rather the universe (also cf. [16]). This entropy transfer is realized by the transport of mass and energy. In doing so, the system can reach a state in which the entropy production is at a minimum and constant with time. Prigogine [7] refers to this state as the *stationary state* of a system, which is characterized by time-independent state variables (including entropy) and constant gradients between system and environment [10, p. 28]. However, Prigogine warns against confusing these *linear stationary non-equilibrium states* of open systems with the (thermodynamic) equilibrium of isolated systems:

> No confusion should arise between such states and equilibrium states which are characterized by zero entropy production [7, p. 75].

In the following, he introduces an example for a stationary system, which is quite suitable for the description of an open geomorphic system, whereas it is confusingly referred to as *equilibrium* in geomorphology:

> Another example of a stationary state is afforded by a system which receives a component M from the outside environment and transforms it through a certain number of intermediate compounds into a final product F which is returned to the external environment.

> A stationary state arises when the concentrations of the intermediate components no longer vary in time. In this case the conditions for the occurrence of a stationary state are expressed by some relations between the reaction rates of the different processes which correspond to the formation or the destruction of the intermediate compounds [7, p. 75].

We thus have to conclude: One of the essential geomorphological definitions of equilibrium (cf. Chap. 6.2), i.e. the equal amounts of inputs and outputs is, from a physical perspective, wrong as it can only be a *stationary non-equilibrium state*.

If the temporal component is taken into account, [7] shows that within *linear* thermodynamic systems the entropy production is reduced with time until it reaches the minimum mentioned above. The law of minimum entropy production expresses a kind of 'inertia'-property of non-equilibrium systems. If the given framework-conditions hinder the system to reach thermodynamic equilibrium (that is, an infinitesimal entropy production) the system moves to the state of 'least dissipation' [13, p. 102].[7]

In other words, this means that the system strives towards that state in which the entropy transfer to the environment is as low as the framework conditions allow. Thus, the stationary state is a stable, predictable minimum of activity that is reached regardless of its initial conditions and which only depends on the framework conditions [17]. Thus, it is an attractor for *linear systems close to thermodynamic equilibrium*.[8] This is an important issue that has to be stressed: This stable, stationary non-equilibrium state can only be reached by systems that behave in a linear manner and that are close to thermodynamic equilibrium. Non-linear systems, that is, all those systems such as river systems, glaciers etc. to which self-organisation is ascribed (cf. Sect. 5.2), do not reach this state.

Just as in the case of equilibrium, all state variables (temperature, entropy etc.) become time-independent in the stationary state, that is, they do not change. Stationary non-equilibrium states are automatically stable [13, p. 104], as the systems are incapable of leaving this state via spontaneous irreversible processes: Internal changes increase entropy production and thus bring the system back to its previous state as soon as a disturbance occurs [7]. This stable state can only be reached in open systems since an entropy transfer to the surrounding is required in order to balance the positive entropy production within the system by a negative entropy flow. Thus, the total temporal entropy variation is zero. This may be achieved, for example, if the entropy of the imported mass is lower than that of the exported mass. The lower the entropy, the higher the proportion of free energy (also exergy, originally called negentropy, cf. e.g. [18]), which in turn is proportional to the existing gradients (e.g. of temperature, chemical concentrations, c.f. [10]). In a sense, the open system 'degrades' the received mass and/or energy to entropy, and due to this degradation it can remain within the stationary state [7]. Thus, it becomes clear that it is not the quantity of inputs and outputs that has to be balanced, but

---

[7] Dissipation means the conversion from one energy form to another, which always happens if work (of systems) is performed (cf. Sect. 5.2).

[8] Here, linearity means the proportionality of flows and forces. Non-linearity, in contrast, is given if e.g. catalysts exist that lead to reaction feedbacks (positive and negative) (cf. [17]).

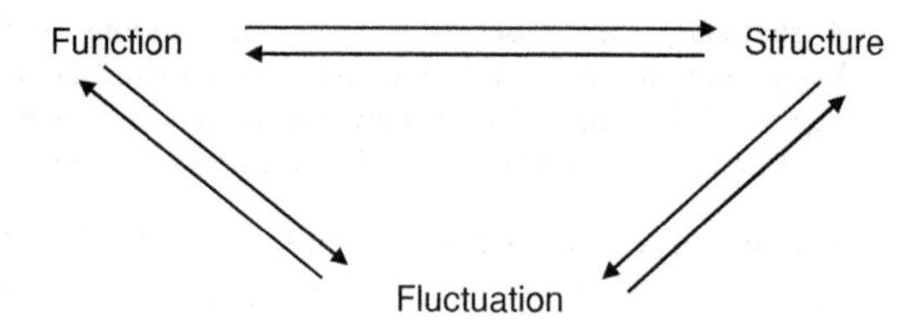

Fig. 5.1 The interplay of the aspects of function, structure, and fluctuation. This interplay can result in 'order through fluctuations' ([14, p. 272], modified)

that the quality of the energy (and mass) flow through the system has to be constant: The uptake of low entropic energy has to be balanced by high entropic energy discharge. These stationary states that follow in some distance to the equilibrium state are also referred to as the 'thermodynamic branch' (cf. [11, p. 12]).

5.2 The Theory of Dissipative Structures

Initially, thermodynamics and especially its second law seemed to be incompatible with our everyday experience, since the mere existence of our structured world with its diverse forms of organisation already contradicts the second law's central statement of increasing entropy and disorder. After all, we do perceive structures everywhere, vortices, sand dunes, stars, galaxies, all of which express macroscopic ordering principles [1, p. 52], also cf. [3].

As these are open systems (which was realized by Prigogine), order or also (self-)organisation are in accordance with physical laws and is the rule rather than the exception [19]. After all, free energy that can be utilized for (self-) organisation, is generated by spatial differences (measured by gradients), and these do not exist in the state of thermodynamic equilibrium but in open systems. If framework conditions are changing in such way that these differences are enhanced, the system is pushed further away from equilibrium and fluctuations form that then build spatial structures (spatial symmetry breaking), or rather spatio–temporal structures.[9] Hence, there are three aspects that are always connected to each other in a dissipative system: the function, the spatio-temporal structure (that results from instabilities), and the fluctuations that cause the instabilities ([14, p. 272], cf. Fig. 5.1).

Structure here means the spatio-temporal arrangement of the system-internal processes [1, p. 66]. A good example is a hurricane: High, threshold-exceeding (threshold here in particular means the instability threshold) gradients in the beginning—e.g. the strong temperature decrease with height—show that the system is far from equilibrium. At the same time, the system is open. Beside these two

[9] The term symmetry-breaking describes the phenomenon when a spatially homogenous situation becomes inhomogeneous, that is, when structures (patterns) form. Thus, for example, patterned ground breaks the spatial symmetry (homogeneity) of the soil, or hurricanes break the spatial homogeneity of the atmosphere (the structures of the hurricane—e.g. the rain bands—are not homogenous to their surrounding).

requirements—the pronounced differences (deviation from equilibrium) and openness—there are thus more prerequisites that have to be given so that the system is able to build spatial structures (such as the eye of the storm and the rain bands): Also the stationary states have to become instable, a situation that is only possible for non-linear systems. After all, only these systems possess the 'instability threshold' that constitute the end of the attractor's dominance close to equilibrium [11, p. 12].

At this point the system exhibits a correlation between system elements on the macroscopic level so that fluctuations can spread from one area to another. For example, coherent movements of molecules (or groups of molecules) are initiated beyond the instability threshold. If these fluctuations, e.g. small convection currents, cannot be supressed by their surroundings, the whole system behaviour spontaneously changes and organises itself into a specific structure (for example, spatial structures such as the rainbands of the hurricane). Beyond this threshold, dissipation is a source of order and re-creation [11, p. 12], also [17]). Non-equilibrium thus virtually is the prerequisite for order and stable structures, just as in geomorphology landforms or specific circulation patterns show. These structures are impossible in a thermodynamic equilibrium.

The irreversibility of real processes that have been introduced by the second law of thermodynamics thus imply a disruption of the temporal symmetry between past and future [1, p. 59], or in the case of non-equilibrium thermodynamics, even a disruption of temporal *and* spatial symmetry [1, p. 60]. Temporal order—in the sense of a before and an afterwards—as well as spatial order arise.

5.2.1 Self-Organisation and Dissipative Structures

But how stable are those states to which the system evolves if it can neither reach equilibrium nor the stationary state of minimum entropy production? Even far from equilibrium a system can evolve towards a stable and steady state, which, however, is no longer characterized by a according potential as, for example, is the case in states close to equilibrium cf. [17]. However, beyond linear thermodynamics the stability is not mandatory anymore as it no longer results from general laws of physics. That means, far from equilibrium system structures are no longer inherently stable [1]. Fluctuations can be increased rather than be regulated downwards as is the case close to equilibrium, and thus these fluctuations can evolve to a completely new order as soon as they capture the whole system. In this case, differences between averages and fluctuations can no longer be defined. The system acts as a 'whole', which is sensitive to the events in all its areas [11, p. 13].

Hence, ordered structures (spatial symmetry breaking) and periodic behaviour may evolve, which is referred to as self-organisation (cf. [11, 20]). However, this does not simply mean the fluctuations of macroscopic parameters such as concentrations, but fluctuations of mechanisms that are changing the (kinetic) behaviour of a system [1], that is, the speed of processes. As fluctuations may lead to a new organisation of the system it becomes clear that there is a close relation between self-organisation on the one hand and the distance to thermodynamic equilibrium on the other hand [17, p. 150]. For example, if a stationary state becomes instable due to changes of framework conditions, from this self-organisation may spontaneously arise, giving the system a specific spatial structure. However,—similar to the biological and sociological second order system theories—this is not synonymous to determination by the environment: In the course of self-organisation, structures and patterns form with different symmetry as it was given by the framework conditions and the external parameters of energy throughput (i.e. symmetry breaking instabilities), a behaviour that profoundly differs from externally driven organisations [19, p. 65].

A common example are the convection cells of the Bénard-instability (cf. amongst others [7]). These structures develop from a gradient, in the case of the Bénard-cells a thermal gradient,[10] whilst these structures, to stay with the example, achieve a much higher dissipation than the conduction and are thus working towards a reduction of exactly these gradients (cf. [10]). On the macroscopic level molecule currents develop that comprise more than 10^{20} molecules [1, p. 52], which is, after all, a highly organised or rather ordered movement. Such evolved structures, for example, are breaking symmetry in the sense that they are horizontally oriented whereas the temperature gradient (that is set by the framework conditions) is vertically oriented [19]. Prigogine (e.g. [7, 13]) calls these structures that arise from fluctuations dissipative in order to stress that there is a close relation between order and structure on the one hand and dissipation and disorder on the other hand. Dissipative structures sustain the entropy production, they 'work' and convert energy [1]. Due to the organisation in structures more entropy is produced, which then is transferred to the surrounding (cf. [10, p. 113])—in line with descriptions of biological systems [1, p. 63] calls this the metabolism of the physical system. After all, self-organisation (just as emergence) needs energetically open systems, so that the excess entropy that is generated within the system can be transferred to the environment (cf. [21, p. 83]).

Self-organisation consequently becomes a characteristic for dissipative structures that are thus differing from equilibrium-structures [1, p. 49]. Another example for structures due to self-organisation are sand-ripples (cf. [22]): The repeating pattern of the sand grains are not determined by the environmental conditions, but the pattern, the structure is set by the internal dynamics of the system. These ordering structures can only be maintained by the flow of energy and mass, thus the emergence of structures and order can now be understood as expression of self-organisation of non-equilibrium-systems, which, furthermore, proceeds dynamically. Thus, what is called structure here is by no means anything

[10] When Bénard found out about those convection cells that are named after him he heated his test liquid (spermaceti) at a room temperature of 20°C with 100°C from below. That means he induced a thermal gradient of 80°C on 1 mm, or 8000°C on 1 cm. However, the convection cells only develop at a specific critical gradient—if it is too low or too high, they do not develop. This can be described by the dimensionless Rayleigh-number (cf. [10], pp. 112–117).

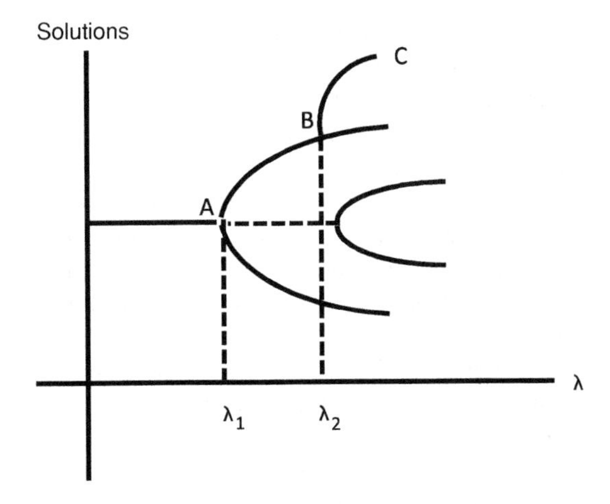

Fig. 5.2 Successive bifurcations: For λ_1 there is only one solution, for λ_2, however, there are several. λ here stands for any characteristic parameter ([14, p. 273] modified)

solid that is made up from the same components throughout, but it is a dynamic regime—it is a *structure of* processes [1, p. 52].

As will be further detailed in Chap. 9.2, this greatly challenges the widespread reductionism: Maybe (!) it is indeed possible to dissect solid structures into ever smaller parts for analysis and to subsequently put them together again—but how should this be possible if the structure consists of processes and thus is constantly changing (cf. [1, p. 54f.])? Furthermore, with the emerging structures, systems become distinct from their environment as they show a coherent activity, e.g. convection, and thus get an individuality towards their environment (cf. [10, p. 111f]). In the context of the previous chapters this seems to be a further potential of the thermodynamic approach: Systems can be distinguished from their environment by their specific structures, they quasi create themselves by giving themselves structures. In this sense, systems could be understood as a giant fluctuation [1, p. 81]. This distinction by means of (process-)structures would at least be a much clearer criteria than the definition of the mechanistic tradition as it is normally applied within geomorphology (cf. Chaps. 3 and 4).

A prerequisite for self-organisation and for dissipative structures are negative and positive feedback loops in which the effect acts back on the cause either in a dampening or enhancing manner (cf. [17, p. 161]). Positive feedback loops such as auto-catalysis act towards instability, whereas negative feedbacks act as stabilisers. In doing so, the system behaviour becomes highly specific, that is, the system behaviour can no longer be derived from a universal law, but each system is a 'particular case' with qualitative distinct behaviour. In contrast, close to equilibrium there is only *one* (stationary) state that only depends on the framework conditions, i.e. some control parameters. If the system is pushed further away from equilibrium, however, at some point it reaches the threshold of instability, that is, a branchpoint or bifurcation point ([17, p. 166], also cf. Fig. 5.2). This can also be described as a kind of crossroad at which the system 'chooses' its system state and its future.

Furthermore, probability theories are also useless at these bifurcation points: The decision, which of the possible future states the system will take happens by chance so that the further system behaviour cannot be predicted. At the same time, due to the fact that the system only choses one of several possibilities, it possesses a kind of 'memory': This one 'choice', this one event affects the whole further development of the system. Therefore, the history of any system becomes crucial for the understanding of present and future system states, since each bifurcation leads to completely different possibilities for the system (also cf. Fig. 5.2). In these zones of instability the fluctuations are decisive for the kind of working regime that the system will then reach [11, p. 14]. This means that the present system state is no longer describable by the inner composition of a system or the framework conditions. The behaviour of systems at the bifurcation points is of special interest, as determination as well as coincidence rule: The fluctuations on the one hand are completely coincidental, just as the 'choice' of the future state is coincidental; however, neither the fact that the system has even reached this point, nor the subsequent path are purely coincidental but depend on the history of the system. Here, coincidence and necessity do not exclude each other but are complementary (cf. [1, 11]).

The processes at the bifurcation points also stress another aspect, namely that of information. Around instability, the system is sensitive for disturbances—be they by internal or external noise, be they by internal or external fluctuations, or be they by purposeful interventions from outside. In the case of hayfork-bifurcation (the system can choose between either an upper or lower branch) the system makes a 1bit-decision as it chooses one of the two branches [19]. This answer to disturbances, however, is selective: from all disturbances the system only filters those that are capable of solving the system-specific branching task. Thus, *the system interprets the disturbances as information within the system context* [19, p. 78]. Thinking further, this means that the system has a bigger room for manoeuvre within its instable phases than within its stable phases, as it can only compensate within the latter case, whereas it can 'choose' from different actions in the first case. This aspect of information due to which a self-organised system may set itself a path to follow shows distinct parallels to the thought of self-reference (Chaps. 3.1 and 3.2): For self-referential just as for self-organised systems the same information (such as a rainfall event) does not prompt the same reaction, but the reaction depends on its internal structure. Jantsch [1, p. 75] goes a step further and claims that physical and chemical systems are also self-referential as they purposefully import and export whatever they need for their own maintenance. However, this view of self-reference substantially differs from definitions presented here and fundamentally reduces the meaning of the concept—that is, that a system only refers to itself in all its actions. Furthermore, the purposeful import and export implies teleology, a purpose of the functioning of a system that I do not share as basic assumption. However, according to the above remarks an understanding of self-reference that is not reduced to the metabolism of a system seems to be also applicable to physical and chemical systems.

Generally, beyond this first bifurcation point the system is considered as 'far from equilibrium' (cf. [10]). From this point onwards, the system can take different steady states,[11] which, however, can be either stable or instable. Which of these states are taken by the system, though, is highly dependent on the structure as well as the history of the system. Furthermore, it can become sensitive to influences that did not play a role in equilibrium (cf. [11, p. 14]). As a consequence of the historicity, of the structural determinacy, and maybe the changed sensibility for influencing factors there is the possibility of same parameter values for different solutions, that is, system states (cf. [20]). This means that the system is indeed adapted to the environment, but that it is not determined by its environment. It means furthermore that the system, to a certain extent, gains autonomy from the environment due to its structural determinacy—a circumstance that might be surprising because of the basic openness of the system. After all, at first sight, this openness gives room for profound influence by the environment. However, it can be directly linked to the autonomy of autopoietic systems as described in Chaps. 3.1 and 4.1. For Prigogine as well as for Maturana structural determinacy[12] is an essential aspect of systems. In both theories it is thus possible that two 'identical' systems with identical environmental conditions take different states. The internal changes are the product of system internal, and not just external, determination. This aspect is central to Maturana: After all, if the state that a system takes was due to the entity with which the system interacts, then the interaction would be an 'instructive' interaction. Systems that are subject to instructive interactions cannot be studied with scientific methods since all instructed systems would take the same state due to identical influence. Hence, they could not be distinguished by the normal observer. After all, two systems can only be distinguished by an observer, because they differ despite influences that are assumed to be identical and they thus are non-instructable systems [23, p. 243].

The temporal path of a system can be physically described by a sequence of stable (deterministic) and unstable (probabilistic) states in the vicinity of the bifurcation points from where onwards the system can follow different future pathways (cf. [17, p. 170]). Here two aspects appear that become crucial for the study of thermodynamic systems, and which, furthermore, may also have far-reaching implications for geomorphology: On the one hand, it is the aspect of historicity—'history matters'. An understanding of thermodynamic systems needs to take the system's history into account. On the other hand, this also implies a difficult step: the abandonment of the claim to be able to predict system behaviour. The same is true for retrospective: conclusions of past system states or of past system behaviour, based on analyses of sediment archives, are not simply self-evident anymore. This limits even further the scope of

[11] In his English translations, Prigogine distinguishes more clearly than in his German translations between stationary states from which a system cannot depart independently, and steady states. The latter are far from equilibrium, but are not stationary, that is, they are not an end-point of a development. I will keep this distinction for logical and linguistic reasons.

[12] I remind again of the rather 'loose' and relaxed usage of the concept of determinism by the authors.

predictions as these often rely on a comparison with past system states. This challenges one crucial and important pillar of geomorphological thinking, as the essential paradigms of the past 100 years or so have been designed to allow for predictions: The erosion cycle of Davis as well as the system theory with its equilibrium concepts are pointing towards certain endpoints of an evolution and thus give the direction of a development, consequently making systems predictable.

5.2.2 *Stability of Structures Versus Stability of Systems*

According to the thermodynamic point of view on system behaviour it can thus be stated that open systems obtain order through fluctuations [17, p. 176]. The fluctuations are maintained by energy and mass flows through the system and, at the same time, structure the system in a way that *might* be very stable against new or further fluctuations, but by no means *has to be* stable. The stability of the structure does not only depend on the amplitude of the fluctuations that it has to resist, but also on the novelty of the fluctuations, that is, fluctuations of values that have not fluctuated before. Thus, no system can be considered as being stable against all possible disturbances (cf. [17, p. 177]).

Especially close to bifurcation points extremely high fluctuations are possible, which can even reach the same magnitude as the averages so that a distinction of fluctuation and average values becomes impossible. Because of the mere size of these extreme fluctuations, correlations between normally independent events may occur—in other words, local events may suddenly affect the whole system. Besides these fluctuations that are triggered by the internal activity of the system, the framework conditions may also show fluctuations, which is why the values that define the interactions can, in most cases, not be seen as constants. As we already know, dissipative systems maintain themselves by the exchange with their environment, which again itself is neither static nor invariant. Thus, the flows between system and environment are constantly fluctuating, and these external fluctuations can also lead to new system structures. Hence, system structures may also display an adaption process to changing framework conditions [17].

However, there is yet another source of instability. These are fluctuations that are caused by the insertion of new components (such as uncontrollable events like technical innovations and mutations). In the case of geomorphic systems these kinds of fluctuations may occur as a result of 'regulating' anthropogenic interventions, as well as the (unintended) insertion of substances that are alien to the system. These new components may then lead to a series of reactions between system elements that compete with the existing functionality of the system [17, p. 183]. Only if the system remains structurally stable against these fluctuations it will not change its functionality and activity.

These thoughts allow for a new point of view within geomorphology: If local events are capable of influencing the whole system, and if these events can even be caused by system-internal processes *without* any external causes, then this has huge implications on e.g. sediment chronologies, since a change of erosion- and/or sedimentation activity cannot be inevitably linked to climate changes, sea level changes etc. In contrast, these findings also have to be also studied with regard to the internal structures of a system (cf. [24, p. 226]). If a system is understood as not being solely determined by environmental factors, the future development of a system depends on the stability of its structures against external and internal fluctuations. The system gains—in contrast to geomorphological system theory—autonomy against its environment: If certain environmental conditions are given, it simply exists, and whether and how it reacts to external changes is highly dependent on its own structure. In this sense, any existing thermodynamic system is also automatically adapted—just as has already been shown for autopoietic systems. If it was not adapted, it would simply not exist (anymore).

With Prigogine's theory of dissipative structures, landforms can thus be understood as a dissipative structure of a system far from equilibrium that might be stable against certain fluctuations. The specific manifestation of a landform, be it a slope or the meandering or braiding of a river course then simply represents a specific dissipative structure that *might be* relatively steady. However, it is clear that this structural manifestation cannot serve as an indicator for an equilibrium state of a system. This is an important difference that—as incisive it is—is not reproduced within geomorphology (cf. Sect. 6).

What does this excursus to physics mean for the understanding of geomorphic systems? Thermodynamics focuses on the macroscopic level, that is, on the system as a whole. Therefore, thermodynamics contrasts mechanics with its focus on the motion of bodies and the relationship between mass, acting forces, and motion. However, the latter *de facto* still is the focus of many geomorphological studies still today: Even if often the respective study object, be it a slope, a river network, or a glacier, is referred to as (thermodynamic) system, nevertheless these studies are based on mechanics. This becomes especially obvious in the case of equilibrium thinking, as will be shown in the following. The problem is that both approaches cannot be reduced to each other. This difference of levels of consideration is often not thoroughly taken into account within geomorphology, thus forming one of the most severe theoretical inconsistencies of geomorphology and severely hampering the scope and value of its scientific statements.

References

1. Jantsch E (1979) Die Selbstorganisation des Universums. Vom Urknall zum menschlichen Geist. Hanser Verlag, Darmstadt, p 464
2. Jantsch E (1987) Erkenntnistheoretische Aspekte der Selbstorganisation natürlicher Systeme. In: Schmitt SJ (ed) Der Diskurs des Radikalen Konstruktivismus. Suhrkamp, Frankfurt/Main, pp 159–191
3. Prigogine I, Stengers I (1993) Das Paradox der Zeit. Zeit, Chaos und Quanten. Piper, München, p 338

4. Kraus H (2000) Die Atmosphäre der Erde. Eine Einführung in die Meteorologie. Vieweg, Wiesbaden, p 470
5. Tipler PA, Mosca G (2006) Physik. Für Wissenschaftler und Ingenieure. Elsevier Spektrum Akademischer Verlag, München, p 1388
6. Iben HK, Schmidt J (1999) Starthilfe Thermodynamik. Teubner, Leipzig, p 108
7. Prigogine I (1967) Introduction to thermodynamics of irreversible processes. Interscience, New York, p 147
8. Graedel TE, Crutzen PJ (1994) Chemie der Atmosphäre. Bedeutung für Klima und Umwelt. Spektrum Akademischer Verlag, Heidelberg, p 511
9. Schlichting HJ (2000) Von der Energieentwertung zur Entropie. Praxis der Naturwissenschaften/Physik 49(2):7–11
10. Schneider ED, Sagan D (2005) Into the cool. Energy flow, thermodynamics, and life. The University of Chicago Press, Chicago & London, p 362
11. Prigogine I, Stengers I (1990) Entwicklung und Irreversibilität. In: Niedersen U, Pohlmann L (eds) Selbstorganisation und Determination. Selbstorganisation. Jahrbuch für Komplexität in den Natur-, Sozial- und Geisteswissenschaften. Duncker and Humblot, Berlin, pp 3–18
12. Rubi JM (2009) Thermodynamik—Wie aus Chaos Ordnung entsteht. Spektrum der Wissenschaft 4(09):30–35
13. Prigogine I (1985) Vom Sein zum Werden. Zeit und Komplexität in den Naturwissenschaften. Piper, München, p 304
14. Prigogine I (1977) Time, structure and fluctuations (The nobel prize in chemistry 1977). In: Frängsmyr T, Forsén S (eds) Nobel lectures, chemistry 1971–1980. World Scientific Publishing, Singapore, pp 263–285
15. Haken H, Wunderlin A (1986) Synergetik: Prozesse der Selbstorganisation in der belebten und unbelebten Natur. In: Andreas D, Hendrichs H, Günter K (eds) Selbstorganisation. Die Entstehung von Ordnung in Natur und Gesellschaft. Piper, München, pp 35–60
16. Baumgärtner S (2005) Thermodynamic models. In: Proops JLR und Safonov P (eds.) Modelling in ecological economics. Edward Elgar, Cheltenham, pp 102–129
17. Prigogine I, Stengers I (1981) Dialog mit der Natur. Neue Wege naturwissenschaftlichen Denkens. Piper, München, p 314
18. Schrödinger E (1999) Was ist Leben? Die lebende Zelle mit den Augen des Physikers betrachtet. Piper, München, p 156
19. Pohlmann L, Niedersen U (1990) Dynamisches Verzweigungsverhalten bei Wachstums- und Evolutionsprozessen. In: Niedersen U, Pohlmann L (eds) Selbstorganisation und Determination. Selbstorganisation. Jahrbuch für Komplexität in den Natur-, Sozial- und Geisteswissenschaften. Duncker and Humblot, Berlin, pp 63–82
20. Nicolis G, Prigogine I (1989) Exploring complexity. An introduction. Piper, München, p 313
21. Mayumi K, Giampietro M (2005) Entropy in ecological economics. In: John LRP, Safonov P (eds) Modelling in ecological economics. Current issues in ecological economics. Edward Elgar, Cheltenham, pp 80–101
22. Merino E, Wang Y (2000) Geochemical self-organization in rocks: occurrences, oberservations, modeling, testing—with emphasis on Agate genesis. In: Krug H-J, Kruhl JH (eds) Nichtgleichgewichtsprozesse und dissipative Strukturen in den Geowissenschaften. Non-Equilibrium Processes and Dissipative Structures in Geoscience. Selbstorganisation. Jahrbuch für Komplexität in den Natur-Sozial- und Geisteswissenschaften. Duncker and Humblot, Berlin, pp 13–46
23. Maturana HR (1982) Biologie der Sprache: die Epistemologie der Realität. In: Maturana HR (ed) Erkennen: Die Organisation und Verkörperung von Wirklichkeit. Ausgewählte Arbeiten zur biologischen Epistemologie. Wissenschaftstheorie, Wissenschaft und Philosophie. Vieweg, Braunschweig/Wiesbaden, pp 236–271
24. Thornes JB (1983) Evolutionary geomorphology. Geography 68(3):225–235

Chapter 6
Fourth Problem Area: Equilibria

From the perspective of physics, geomorphic systems are thermodynamic systems. As they are open, they cannot reach thermodynamic equilibrium as long as there any disequilibrium with the environment exists. Still, the state of equilibrium and thus the thought of steady stability builds a reference model for the problem of describing a big ensemble of interconnected components. However, in doing so, it is often neglected that the equilibrium with its insensitivity of the average values against any fluctuations is only a special, singular case [1, p. 12f]. If the results from Chap. 5 are taken seriously, it becomes clear that the framework conditions play a bigger (close to equilibrium and on the thermodynamic branch) or lesser (far from equilibrium beyond the instability threshold) role in determining the system behaviour, that is, its functioning. Thus, due to the basic openness of geomorphic systems, we can at best assume stationary states within which entropy production is minimal and is compensated by a negative flow of entropy (that is, entropy is transferred to the environment) (cf. [2, p. 78]). However, it is a precondition for these stationary or stable states that the system is close to thermodynamic equilibrium and linear. Due to the surplus of free energy on Earth, this, however, is not true for most geomorphic systems. The American geologist and 'father of equilibrium thinking', Grove Karl Gilbert put this in words more than 130 years ago:

> It is evident that if steep slopes are worn more rapidly than gentle, the tendency is to abolish all differences of slope and produce uniformity. The law of uniform slope thus opposes diversity of topography, and if not complemented by other laws, would reduce all drainage basins to plains. *But in reality it is never free to work out its full results; for it demands a uniformity of conditions which nowhere exists.* [...] The reliefs of the landscape exemplify other laws, and the law of uniform slopes is merely the *conservative element* which limits their results ([3, p. 115], accentuation by KvE).

Despite Gilbert's point of view that an equilibrium state simply is *one* amongst many system states and which is thus nothing that a system inevitably needs to reach (also cf. [4]), equilibrium thinking, especially in analogy to the mechanical

K. von Elverfeldt, *System Theory in Geomorphology*,
Springer Theses, DOI: 10.1007/978-94-007-2822-6_6,

equilibria as described above, has been dominant within geomorphology at least till the 1990s and still influences a considerable amount of geomorphological studies:

> Equilibrium continues to be referred to in a wealth of research publications [...] and [...] many others include implicit references to the concept [5, p. 167].

Prigogine was apprehensive of a confusion of stable states and equilibria, and this is indeed what happened in geomorphology: neither is there a consistent definition nor does a consistent usage of the term 'equilibrium'exist, and furthermore there is no clear distinction between stable, stationary and equilibria. The concept of equilibrium is central to geomorphology and was and is one of the fundamental basics of process geomorphology ([6], also cf. [7]). These issues open up a problem horizon that needs clarification.

6.1 After All, what is 'Normal'? On Equilibria as 'Normal State'

The idea that nature holds itself in a kind of balance or stability is neither new nor restricted to geomorphology (cf. [8]). Here,'equal' means that a strategy is acting that dampens all efforts to overcome the established state [9, p. 35]. Traditionally, nature is viewed as being in equilibrium when no changes, or later only small deviations of a 'normal state' can be observed, e.g. of landforms or of system behaviour. Generally, the observation period, that is, in most cases several years or some decades, builds the reference frame. Firstly, this is due to the fact that there are simply no further reaching data series, but secondly this is also due to the fact that Man unknowingly transfers his own horizon of experience to other fields [10, p. 7], so that his experiences easily and unwittingly become the measure for observation (cf. [11]). Within this perception of 'normality' disruptions of equilibrium are caused by disturbances that lead to measurable changes within the system. Increasingly, Man was and is identified as being a (main) disturbing factor, which often leads to the conclusion that nature would be in equilibrium if mankind did not interfere (e.g. in [12, 13]; also cf. [8]). Furthermore, this perspective implies that a system reaches or at least strives towards a new equilibrium state after every disturbance. Implicitly or explicitly, the equilibrium state thus becomes a normal, ideal, or pursued state.

Within equilibrium, any single event or any single system element does not play an important role for the behaviour of the system or for the respective system state: If the system is big enough (i.e. if it consists of a sufficient amount of elements) the existence or non-existence of a single element is as unimportant as a single blade of grass within a meadow. The opposite is true for the situation of non-equilibrium within which the single object becomes significant (cf. [9, p. 35], as well as [14, p. 35]), since the individual behaviour of any single element can be a triggering factor for self-organisation.

As already mentioned, within geomorphology (and especially within English literature) the concept of equilibrium is generally traced back to Grove Karl Gilbert (1843–1918) ([3, 15]; e.g. cf. [5]; Ritter et al. [7]). His study from 1877 focussed on the description of the geology of the Henry Mountains. In the fifth chapter of his report (page 99ff), Gilbert addresses the landforms of his study area and their forming forces and ponders on how the interplay of (erosive) forces and the (resistive) bedrock leads to exactly those forms he finds. He proposes an equilibrium of action by which he explains the temporal development of slopes and the river bed. Gilbert [3] argues that a river balances slope differences by increased or decreased bed erosion until the erosion capacity is proportional to the resistivity of the bedrock everywhere. He refers to this state as "equilibrium of action" [3, p. 113], or, synonymously, as "dynamic equilibrium" [3, p. 123]: This equilibrium concept can be seen as widely analogous to the mechanical equilibrium of forces described above (also cf. [16]).

According to Gilbert, changes within the system are due to changes within framework conditions (climate, tectonics, ...) to which the system successively adapts. Here, Gilbert especially stresses the interdependence between single system components:

> The disturbance which has been transferred from one member of the series to the two which adjoin it, is by them transmitted to others, and does not cease until it has reached the confines of the drainage basin. For in each basin all lines of drainage unite in a main line, and a disturbance upon any line is communicated through it to the main line and thence to every tributary. And as any member of the system may influence all the others, so each member is influenced by every other. There is an interdependence throughout the system [3, p. 124].

Due to the dominance of the linear, landscape-evolution focussed concept of the geographical cycle by W. M. Davis [17], Gilbert's ideas of a process oriented geomorphology were forgotten throughout the first half of the twentieth century Gilberts (also cf. [18]).[1] Indeed, Davis [17] also utilized an equilibrium concept, but this referred to the relative age of landforms: graded rivers were thought to be associated with the stadium of maturity. It was only in the middle of the 1940s that the process-oriented concepts of Gilbert were re-discovered (cf. e.g. [19, 20]), and also the Davisian concept of rivers at grade was re-evaluated with regard to the 'new' perspective. One of the key differences between the equilibrium concept of Gilbert and nearly all of the subsequent approaches is that Gilbert never assumed that a system is at any point of time in a specific or 'normal' state (cf. [4]). How equilibrium has changed since Gilbert will be

[1] According to Orme [18], cyclic concepts were a typical characteristic of scientific thinking throughout the nineteenth century ("cycle mania", [18, p. 475]). The simple and seemingly logical concept of Davis was attractive and persisted relatively unchanged until the middle of the twentieth century. It was new insights into plate tectonics, new dating methods, as well as a new understanding of climate changes and changing surface processes that caused a renunciation of the Davisian model.

shown in the following by means of a chronological overview of essential works on geomorphic equilibria.

6.2 A Historical Overview of the Geomorphological Equilibrium Concept

In fact, a steady state must be rare.

Hack [21]

In his work, Horton [19] focussed on the relation between process and form and stressed the mutual adjustment of both, which could finally lead to an equilibrium. Horton understood equilibrium as a balanced erosion and transport capacity so that no further erosion occurs. Insofar, it is not synonymous to Gilbert's dynamic equilibrium that states that the erosion capacity is proportional to the resistance of the bedrock. Furthermore, also Mackin's [20, p. 471] concept of graded rivers is not synonymous to Gilbert's dynamic equilibrium:

> A graded stream is one in which, over a period of years, slope is delicately adjusted to provide, with available discharge and the prevailing channel characteristics, just the velocity required for transportation of the load supplied from the drainage basin. The graded stream is a system in equilibrium; its diagnostic characteristic is that any change in any of the controlling factors will cause a displacement of the equilibrium in a direction that will tend to absorb the effect of the change.

Interestingly, Mackin explicitly refers to a temporal scale within which the graded river should be studied. Herewith, on the one hand seasonal or temporary fluctuations, and on the other hand slow changes that accompany Davis' erosional cycle should be excluded. Consequently, Mackin suggested several years as "natural unit" [20, p. 477]. He suggested furthermore that long-term balance is the distinctive characteristic of a graded river. Mackin himself considered his work to be a revision of Gilbert's and Davis'equilibrium concepts. By following up the equilibrium concept he explicitly disagreed with John Kesseli [22, p. 561] who stated:

> The concept of the river at grade assumes the attainment of a balance between power and load. A critical review of the original sources responsible for this notion of balance in stream work, particularly A. Surell and G.K. Gilbert, reveals that this equilibrium is only a postulate which remains unsupported by observations and that the attributes commonly associated with a graded river, like absence of downward corrosion and efficiency in lateral erosion, are incompatible with the postulate of balance between power and load. Consideration of the common variations in volume of flow and of the importance of turbulence in transportation of debris suggests impossibility of maintenance and even of acquisition of a full load.

In contrast, Mackin [20] considered the equilibrium concept as a basis for the modern, quantitative description of river transport.[2]

Two years after Mackin, Strahler [23, 24] published the first and the second part of his studies on the theory of equilibrium of erosional slopes. He aimed at relating causing factors and resulting form characteristics quantitatively to each other, using the example of erosional slopes:

> Under equilibrium theory, [...] slopes maintain an equilibrium angle proportional to the channel gradients of the drainage system and are so adjusted as to permit a steady state to be maintained by the process of erosion and transportation under prevailing conditions of climate, vegetation, soils, bedrock and initial relief or stage. Thus, both slopes and streams are graded [24, p. 810].[3]

According to his initial thesis that he assumed to be confirmed by his quantitative analyses, slopes (of a lithological and relief energetical uniform region) mirror a steady state[4] of erosion and input of material. Strahler [23, p. 676] states:

> The concept of equilibrium has long been applied to graded streams and their associated slopes but the nature of this equilibrium and its basic similarities with other systems of equilibrium in nature seem not to have been fully examined.

Strahler describes a graded river as open system that consumes energy in order to maintain its steady state. If the energy or mass flow is disturbed, the open system will adapt itself until a (new, different) time-independent, steady state is established. The steady state manifests itself by the moulding of certain, time-independent state of form characteristics, so-called "equilibrium forms" [23, p. 676]. At the same time, erosion and transport processes reach a steady and uniform flow—averaged over several years or decades. But the system is also capable of self-regulation: Open systems can adapt themselves internally to changed energy and mass inputs and thus reach a new steady state [25]. Observed on longer timescales, however, (equilibrium) forms show a slow evolution as they continuously adapt to changed geological framework conditions. For Strahler [23], the advantage of the steady state concept was the new focus on the interplay of dynamics and morphology, thereby allowing a quantitative assessment of typical

---

[2] Mackin's own work of 1948, however, was qualitative, as he stressed himself.

[3] Apparently, Strahler uses the term equilibrium for landforms (e.g. the slope angle), and the term steady state for processes. Grade, in turn, seems to express that in a system there is a steady state as well as equilibrium. In contrast, Mackin uses grade and equilibrium synonymously. In physics, again, steady state describes something completely different: a steady state of a system far away from (thermodynamic) equilibrium, and not the constancy of processes.

[4] Within German speaking geomorphology, steady state is mostly translated as 'stationary state' (*stationärer Zustand*). However, the term 'stationary' implies a certain invariableness, immobility, and inflexibility, which is not coherent with the descriptions of the state, and which, moreover, within physics means closeness to thermodynamic equilibrium that is not intended within (German) geomorphological usage at all. Therefore, in the following I will continue to call these states steady in order to make clear that a deviation of these states is basically possible anytime. However, the conflicting usage with that of physics cannot be avoided thereby.

forms.[5] However, this perspective is problematic as, for example, two, and strictly speaking even three different and conflicting theoretical approaches were mixed by Strahler [23, 24] and later on also Hack [6]: Gilbert's [3] concept of dynamic equilibrium, on the one hand, is anchored within the physics of mechanics, whilst it is at the same time a geomorphologically adapted concept (cf. [16]). By referring to a steady state, Strahler [23, 24] and many subsequent authors (very obvious e.g. in [26], but also in [27]) further mix this mechanistic-geomorphological concept with a thermodynamic concept, which is, however, based on completely different and incompatible basic assumptions within statistical physics (also cf. [16]). One could say that latest from this time the foundation of wide parts of process geomorphology became—from the viewpoint of physics—an inadmissible and half-baked mixture of two physical theories that are valid for completely different scientific areas.

The German geomorphologist Frank Ahnert occupied himself for several decades with questions of (dynamic) equilibrium and of steady state [26, 28–31]. Moreover, with Ahnert [28] one of the reasons for the attraction of the equilibrium concept can be illuminated: According to Ahnert, systems 'try' to reach a specific state, that is, they quasi have an aim as the increase in debris input *tries to compensate* the output of debris and *to establish* a dynamic equilibrium between the two [28, p. 62]. Due to this teleology systems appear to be controllable and less complex, as key variables are seemingly obvious. These aspects are basic to any steering attempts towards systems. Ahnert assumes that Gilbert's [3] dynamic equilibrium is synonymous to Hack's steady state ([6], see below) and proposes that a steady state with constant relief is extremely unlikely as this would need a very improbable constancy of framework conditions [29]. Since a steady state relief cannot be reached, Ahnert [29] focuses instead on the intrinsic striving towards dynamic equilibrium as well as the time a system needs to react and adapt to changes (relaxation time) [28–31]. In one of his later works, Ahnert [26, p. 138] eventually describes the tendency towards a dynamic equilibrium as a universal principle within process-response systems, regardless of the question whether the dynamic equilibrium really exists.

In 1957, Culling [32], as well as Mackin [20], tried to apply Strahler's [23, 24] concept of open systems to river systems, especially graded rivers, albeit still in the framework of cycle thinking. Culling pursues the extent to which open systems may reach steady and thus time-independent states as postulated by classical thermodynamics. Culling [32] notes that, in consequence, energy values should be studied and not mass transport, as is usually done. However, as the result of the impossibility to operationalize an energetic approach, he also uses mass transport, in his case bedload transport. He points out that within steady state, any river segment has to fulfill the condition of invariant mass transport through this segment with time [32]. Since it is an open system, changes within the external conditions also

[5] Strahler [23, p. 676] calls these typical forms "*equilibrium forms*" as well as "*forms typical for the steady state*". Whether he differentiates between the terms is not obvious.

influence system behaviour, however, the system tends to dampen or counteract these disturbances (LeChâtelier-principle, also cf. [33]) and thus thrives towards a steady state.[6] This balance of disturbances is only successful, however, as long as the external impacts do not exceed certain critical values in magnitude or speed of onset. Culling concludes that graded streams can indeed reach steady states, but they do not necessarily need to. At that time and the prevalent equilibria mania, these thoughts can be interpreted as rather explicit criticism of the equilibrium approach, however, without fundamentally challenging it. 30 years later, though, Culling is much more explicit and stresses that linear approaches are too trivial and restrictive to depict the manifold complexity of the world (cf. [34, p. 57]). Furthermore, Culling states that with their system approach, for a long time geomorphologists held a naive attitude in approaching the world [34, p. 68].

Eventually, Hack [6] explicitly introduces the equilibrium and system approach as alternative to the Davisian erosion cycles. Furthermore, he regards it as being the philosophical foundation for the analysis of the interplay of process and form—thus, the farewell of the Davisian model proceeded quickly and vehemently in the second half of the twentieth century. Hack perceived the landscape and its moulding processes as components of an open system in a steady state of balance within which any slope and any landform adjust to each other—a concept that he adapted from Grove Karl Gilbert (see above). However, it has to be stressed that Hack's dynamic equilibrium substantially differs from that of Gilbert: Hack intended to explain long-term structures of the landscape by short-term processes (cf. [35]). Although Hack—as well as Ahnert later on—assumes that the tendency towards a dynamic equilibrium is a "universal principle" [21, p. 87], in his eyes this does not inevitably imply a steady state.

Just as Gilbert [3], Hack [6] assumes that within an erosional system all topographic elements are eroded at the same rate. According to Hack, by means of dynamic equilibrium the topography as well as the differences between landforms can be explained independently of time, whereby spatial correlations could be focussed.

Twelve years after Strahler's first attempt to transfer the theory of open systems [36, 37] to geomorphological issues, Richard Chorley [38] published his paper on geomorphology and general system theory. He suggests that the erosional cycle of Davis [17] can be seen as the geomorphological equivalent to the second law of thermodynamics for isolated systems in some sense (cf. [38, p. B2f]). This he contrasts by the theory of open systems as propagated by Ludwig von Bertalanffy and transfers it to the study objects of geomorphology, such as catchments, river sections, and other landforms. Furthermore, Chorley refers to Bertalanffy's concept of steady state that—looking back and knowing the theory of dissipative structures—more or less is equivalent to Prigogine's stationary state of

[6] Here, it becomes obvious that Culling assumes—congruent to his time—that systems are linear and close to equilibrium. At the same time, his description is equivalent to descriptions of the stationary state of linear thermodynamic systems within physics.

non-equilibrium systems.[7] Bertalanffy (e.g. [36, 37, 39]) as well as Chorley [38] understand the concept of steady state to be synonymous to that of dynamic equilibrium.[8] Characteristically, the system brings itself back to equilibrium by means of self-regulating processes. Also, Chorley ([38], p. B5) assumes that open systems inevitably strive towards a steady state:

> An open system will, certain conditions presupposed, develop toward a steady state and therefore undergo changes in this process. [...] The trend toward, and the development of, a steady state demands not an equation of force and resistance over the landscape, but that the forms within the landscape are so regulated that the resistance presented by the surface at any point is proportionate to the stress applied to it.

As soon as the steady state is reached, the initial situation as well as the history of the system loses importance.[9] Furthermore, Chorley [38] assumes that geomorphic systems are capable of building order due to their openness. As a consequence, (spatial separating) structures develop. However, Chorley had to admit himself that this contradicts the afore stated mutual adjustment of acting forces.

In 1965, Schumm and Lichty [40] published a paper with the aim of assigning a concrete temporal frame to the thoughts of Hack [6] and Chorley [38] in order to dispel the impression that the new system approach implies a complete negation of the temporal dimensions of geomorphic systems. Rather, it just means a focus on much shorter timescales. They emphasize that geomorphic systems should not be analysed without their history and further explain:

> We believe that distinctions between cause and effect in the moulding of landforms depend on the span of time involved and on the size of the geomorphic system under consideration. Indeed, as the dimensions of time and space change, cause-effect relationships may be obscured or even reversed, and the system itself may be described differently [40, p. 110].

In other words: Only if the system is assumed to be static, can cause and effect (and consequential possible equilibria) be detected at all. If a dynamic approach is chosen, any cause can become an effect and any effect can become a cause [40, p. 113]; consequently, the system no longer shows linear behaviour and it is impossible to state an equilibrium of a whole system: "A choice must be made whether only components of a landscape are to be considered or whether the system is to be considered as a whole" [40, p. 112]. After all, only if the correlations of single variables are analysed, can time or the history of the system be

[7] However, despite the explicit reference, Chorley's understanding of the steady state differed fundamentally from that of Bertalanffy: According to Chorley, the import and export of mass and energy is held constant because the system changes its form or geometry, which is something that is impossible in Bertalanffy's steady state since the system regulates itself back to its initial state.

[8] Bertalanffy [39] suggests "Fliessgleichgewicht" as translation for "steady state". However, I think that this term is highly ambiguous and thus not useful. In the German (original) version of my thesis I consequently avoided this term.

[9] At this point it should be emphasized again, that according to Prigogine this is a characteristic of linear stationary non-equilibrium systems close to thermodynamic equilibrium.

neglected. It is exactly this approach that has been chosen for several years (cf. below [41] and [42], representative for many). Schumm and Lichty [40] illustrate that a temporal dimension is connected with the spatial extension of any geomorphic system and vice versa. In order to reconcile the historical and process-oriented perspective they suggest a concept of scale within which they define the different timeframes cyclic time, graded time, and steady time. For better comprehension, Schumm and Lichty [40] reduce their hypothetical catchment area to a longitudinal river profile. The study of Schumm and Lichty [40] did not only largely influence geomorphology as a whole, but is also one of the reasons why the equilibrium concept survived for so long within geomorphology. However, the adding of a temporal (and spatial) scale for (different) equilibrium states is another geomorphological distinctiveness, thereby enhancing the development that certain physical concepts within geomorphology are neither congruent with their original physical definition, nor with the usage in other scientific disciplines (also cf. [16]).

Only two months later, Howard [41, p. 302] also stresses that the history inherent to landscapes, and the striving for equilibria are not contradictory to each other, and that both concepts can be unified within the system approach. How strongly past events influence the universal tendency towards equilibrium depends, according to Howard [41], first, on the intensity of the respective events, and second, on the time that has passed since then. He describes geomorphic systems as passive, that is, changes within the system only occur due to changes of external variables. This is explicitly meant in a deterministic way: Any (not random) alteration within the system can be assigned to a change of an external variable. When an external variable changes, the system reacts as an organic whole, that is, all system parameters are re-adjusted [41, p. 303]. For Howard [41, p. 305], equilibrium thus is “a complete adjustment of the internal variables to external conditions”. Whether an internal variable is in equilibrium can be assessed by the constancy of this variable. Another characteristic of equilibrium is maximum erosion efficiency. For Howard, the only possibility that a system is in long-term metastable disequilibrium is the existence of thresholds [41, p. 308]: If external variables induce a system or parts of it to cross thresholds, the change might be too substantial to return to equilibrium. In 1982, Howard further elaborates his 20 year old thoughts with the aim of presenting a quantitative definition of equilibrium. He starts his paper with a sharp critique of the inconsistent, inaccurate, overlapping, and conflicting definitions of equilibrium within geomorphology. In contrast to his earlier paper, Howard [43, p. 303] defines equilibrium quantitatively by a

> temporally invariant relationship between the values of an output variable and the values of the input variable(s) in a geomorphic system. Disequilibrium occurs if the output deviates from the functional relationship by more than a consensual degree.

Howard clarifies that equilibrium is no characteristic of a whole system, but that it always refers to one or more input and output variables. As he, consequently, cannot and does not want to make any statements on the structure of a system, Howard [43] calls the states as described by him equilibrium and explicitly not steady state, as he wants to avoid the thermodynamic connotation.

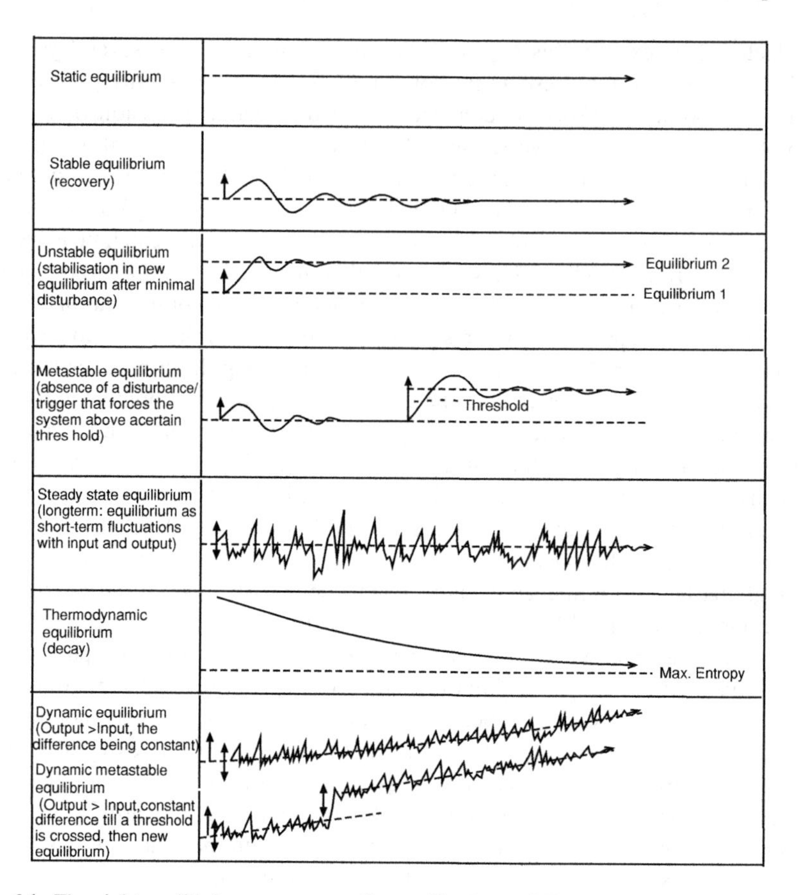

Fig. 6.1 The eight equilibrium types according to Chorley and Kennedy [44], p. 202, modified

In 1971, Chorley and Kennedy presented eight equilibrium types (cf. Fig. 6.1) on almost 50 pages. Dynamic equilibrium they define such that it shows a clear trend around which short-term system states oscillate, and they further distinguish a subtype, i.e. the metastable dynamic equilibrium [44, p. 202]. In the understanding of Chorley and Kennedy [44], dynamic equilibrium is thus no longer equivalent to a time-independent (steady) state. Generally, both in physics and in geomorphology, however, the existence of a trend is just seen as evidence for non-(or dis-)equilibrium—a criticism that is especially also true for their illustration of the thermodynamic equilibrium [44, p. 202]. Gilbert's [3], Ahnert's [26] and others' ' dynamic equilibrium' corresponds with Chorley's and Kennedy's [44, p. 202] "steady state equilibrium". What is defined as "static equilibrium" by Chorley and Kennedy [44] is equivalent to the thermodynamic equilibrium, whilst stable equilibrium basically is a description of the stationary state in thermodynamics, or circumscribes the principle of LeChâtelier.

The idea that within equilibrium, processes reach optimal efficiency [44, p. 204] is in complete contrast to the thermodynamic understanding of equilibrium.

Chorley and Kennedy [44, p. 205] even go a step further: Only if the system fulfils any of the conditions of equilibrium can it even be regarded as an entity—equilibrium mutates to the final definition criterion of geomorphic systems.

The two central aspects of predictability and teleology illuminate why the system approach became so pivotal to geomorphology. Another important issue is the causal linkage of in- and outputs, as can be exemplified by the 1975-paper by Stanley Trimble [45]: He stated that since the settlement by Europeans most US-American rivers have not been in steady state.[10] However, it becomes obvious that Trimble assumes a previous system state as a kind of 'ideal case' or 'normal state'(cf. also Chap. 6.1) from which it now deviates. According to him, the deposition of colluvia and alluvia are an indicator for "a definitive lack of the steady state" [45, p. 1208]. From this an analytical problem arises: As the system is not in a steady state, no simple causal relation between denudation rates in the catchment and sediment yield can be assumed. The importance of equilibrium thinking for geomorphology thus is due to the fact that it allows for linear causal relations and direct inference from one variable to the other. For example, this is also expressed by [7, p. 6]:

> Dynamic equilibrium suggests that elements of landscape rapidly adjust to the processes operating on the geology, and thus process and form reveal a cause and effect relationship. The forms within a landscape maintain their character as long as the fundamental controls do not change.

For Schumm [46], the question of the functioning of a landscape is central to his research. Thus, according to him it is necessary to focus on short timespans, to proceed actualistically, and to avoid extrapolations of results on longer timescales. In doing so, Schumm [46] pursued several aims: First, he wanted to counter the paradigm of the progressive change of landforms, second, he aimed at focussing on the details of the landscapes. Consequently, neither the conceptual frame of historicity (Davis) nor that of the dynamic equilibrium were useful. With his focus on applied and problem-oriented research, the equilibrium concept was of low scientific interest as it cannot explain discontinuities or change. Furthermore, he considered the assumption unrealistic that changes within the system should only be attributable to an alteration of external conditions (humans, climate, geology). In addition, Schumm [46] does not consider as given that within dynamic equilibrium all landscape elements react in the same way. Unfortunately however, Schumm does not detail to which of the equilibrium concepts he refers, but he seems to have Strahler [23, 24] and Chorley [38] in mind. Schumm [46] states that those approaches are not helpful for his research and proposes the concept of thresholds, complemented by intrinsic and extrinsic thresholds. Still, Schumm's criticism of the equilibrium concept was not general, but only directed to his very specific research focus—on longer timescales he indeed assumes that it is

[10] Trimble was immediately criticized by Schumm et al. (1976) for applying the concept of steady state to a wrong timescale. Therefore, according to Schumm et al. (1976) no other conclusion has been possible than that the system is not in steady state.

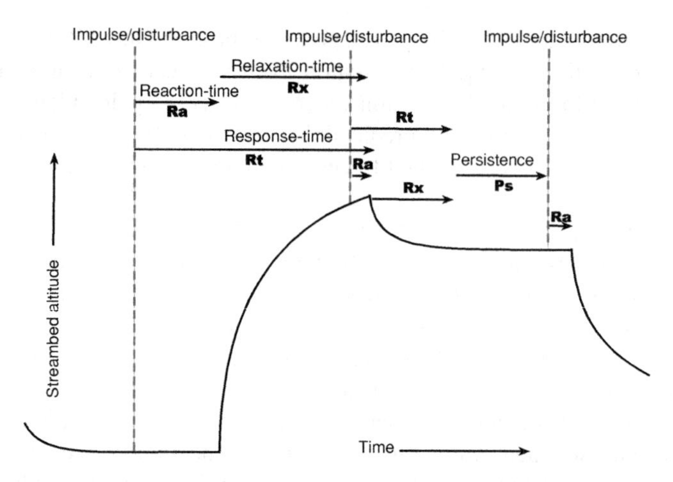

Fig. 6.2 Scheme of the reaction of a geomorphic system to external disturbances, exemplified by streambed altitude (from: [47, p. 133, 42], modified)

compatible with the assumption of thresholds. After all, the existence of a threshold does not generally hinder equilibrium, but may simply delay it.

Eventually, it is exactly this approach of both thresholds and equilibria that is utilized by Bull [42] to describe landscape changes (cf. Fig. 6.2). Equilibrium for him means the balance between those processes that are, for example, acting in a fluvial system, or if put in a diagnostic way, the absence of deposition or erosion (for river systems), and an unchanged configuration of the landscape (in the case of slopes). When a system is disturbed the system only reacts to this disturbance if thresholds have been crossed. If a threshold is crossed, this includes a change in the mode of operation—for example, just an increased deposition is no indication for threshold-crossing. For illustration purposes, Bull [42] uses the example of streambed altitude: Disturbances cause a change of streambed altitude. Thereby, the system undergoes different phases (cf. Fig. 6.2): After the initial disturbance, the system is in the phase of reaction (Ra), then it reaches the phases of relaxation (Rx) and persistence (Ps). The first two phases, that is, that of reaction and relaxation time, are generally taken together as response time. As soon as the information about the disturbance has been processed within the system, the streambed altitude starts to change. If the system is disturbed before the new equilibrium has been reached, the system cannot balance the changed framework conditions but begins to react to the new disturbance.

Just as in Trimble's study [45], this example shows that equilibrium is understood as a steady and normal state within which resisting and driving forces are balanced (cf. [47], also cf. Elverfeldt and Keiler [48]). Persistence time shows how long a new equilibrium will be maintained. However, Ritter et al. [7, p. 18] point towards problems of the application of the threshold concept as suggested by e.g. Bull [42]:

> ... applying the threshold concept in real situations is difficult and is often complicated by failure to consider the factor of time.

In contrast to Bull [42] they assume that a threshold is only crossed when the system subsequently reaches a new equilibrium:

> The real question, then, is whether the disruption of equilibrium produces landform or process changes that are meaningful in geomorphic analyses ...; in other words, what is temporary disequilibrium and what is geomorphologically significant disequilibrium? [7, p. 18].

The listed examples, some of which are much-quoted within geomorphological literature, show that equilibrium thinking persisted for several decades—and surprisingly even also after a change towards approaches of non-linearity. The endurance of equilibrium thinking is even more remarkable, as restrained as well as marked criticism of the equilibrium approach occurred quite early on. (cf. e.g. [22]). Also Bremer [27] points towards the difficulties of actually assessing equilibrium, as e.g. balanced incision and deposition is not necessarily mirrored in the geometry of a river—an observation that was repeated by Phillips [49] about twenty years later. The equality of inputs and outputs can also not be considered as equilibrium condition of e.g. a river system (c.f. e.g. [50]). Bremer [27, p. 13] finally concludes that the idea of equilibrium is not very useful for the assessment of river morphology. According to her, the lacking universal validity of the equilibrium approach is due to the omission of different climate zones, and of inherited traits within the concept.

6.3 Criticism of Equilibrium

> *Geomorphologists, unfettered by data, have wandered through the equilibrium marketplace, shopped whimsically, and purchased for themselves a veritable Pandora's box.*
>
> Thorn and Welford [16]

Both conflicting definitions and the synonymous use of terms such as "dynamic equilibrium", "quasi-equilibrium", and "steady state" contributed considerably to the eventually increasing discontent with the concept (e.g. [51, 52], also cf. [16]). Already as early as 1970, Young stated that there are 14 different and partly conflicting equilibrium concepts within geomorphology [53, p. 586ff]—a list that probably was not complete back then and would have encountered several additions ever since (cf. [26]). In 2007, Phillips presented a list with even 23 different definitions from geomorphology and relevant neighbouring disciplines [49, p. 7ff].

Montgomery [54, p. 47] also emphasizes the inconsistencies of geomorphological equilibrium definitions within the discipline, but does so in comparison to definitions within math, mechanics, and thermodynamics and complains about the semantic laxity of the definitions. Thorn and Welford [16] point in the same direction when revealing the variety of definitions of dynamic equilibrium. They suggest replacing this ambiguous term by the—according to them—clearer concept of mass flux equilibrium [16, p. 686]. Ahnert [26], on the other hand, tries to introduce a purely geomorphological distinction of dynamic equilibrium and steady state. He stresses that steady state refers to the system as a whole (no netchange of the components) whilst dynamic equilibrium refers to the interplay of processes and their respective rates. Ahnert [26] does not follow the proposition of Schumm and Lichty [40] of temporal differentiation but rather attempts to reduce the variety of equilibrium terms to the two afore mentioned.

In a paper on the ecological stability concept, the results of which can be widely transferred to the geomorphological equilibrium concept, McCoy and Shrader-Frechette [8, p. 192f] conclude with several reasons why it is not expedient to further follow equilibrium approaches. One substantial issue, also fully valid for geomorphology, reads: the variety of meanings of the term "equilibrium" is rooted in conceptional incoherency, and even inconsistency. Different definitions refer to different aspects and often do not even describe system characteristics; some refer to the time during which changes occur and not to the changes themself; and others assume different temporal and spatial scales than the next. For example, for Ahnert [26] the difference between steady states and equilibrium is that the first focuses on the system as a whole, whilst the second focuses on the processes. Schumm and Lichty [40] as well as Chorley and Kennedy [44], on the other hand, assume that the difference is in the respective spatial and temporal scale. So far, I was neither able to find a single universally applicable specification that describes what exactly characterizes equilibrium, nor any information on sufficient or necessary conditions that have to be met so that a system can be assumed to be in equilibrium. The consequence is: Due to its inconsistency and ambiguity the concept of equilibrium hampers scientific progress rather than promoting it.

Furthermore, there are more problems attached to the concept of equilibrium than simply that of incoherent definition. For example, the idea of system equilibrium can only be maintained if system analysis is restricted to one single variable in a specific time unit (in Bull's [42] example this was streambed altitude, also cf. [43]). Thus, the equilibrium state of a whole system as generally understood within geomorphology at best is a local, in most cases: mechanical equilibrium of single variables. This can indeed be useful for some research questions, however, it says nothing about the system (state) as a whole. Yet, it is exactly such a focus on single elements that has been criticized by Chorley and Kennedy [44] as not being helpful and thus was one of the reasons why they introduced systems theory to geomorphology as they intended to replace this focus. Obviously, they did not succeed, as the focus remains on single compartments:

> The balance or equilibrium within our system is revealed by statistical relationships between the various parameters; for example, basin area may be directly related to total channel length and so on [7, p. 6].

Only this assessment of single variables allows at all for observing equilibria—but these are only equilibria of single variables. Still, based on this assessment of a single parameter the equilibrium of the whole system is implied. What we miss, however, is that this equilibrium does not exist anymore from the very moment of disturbance—due to the focus on one variable we simply do not perceive the imbalance. What is described here as an equilibrium of a system, thus should rather be regarded as a state description of a single variable with its measurable or perceivable reactions to changes. Therefore, the assessment of components is completely contrary to the holistic claim of system theory. This is especially noteworthy as in geomorphology, too, system theory is and was understood and propagated as alternative draft to reductionism (cf. [44, 55, 56]).

Even more serious than the inconsistencies and the open contradiction to the general system approach is the problem that the usage of the equilibrium and state terms within geomorphology are diametric to the definitions in thermodynamics. This is one of the reasons why Thorn and Welford [16, p. 666] conclude: "Equilibrium is more nearly a source of confusion than of enlightenment in geomorphology". More pointed: Equilibrium thinking is Pandora's box for geomorphology [16, p. 692]. As physics is one of the main providers for new impulses—for example, also within geomorphology some research objects can be considered as thermodynamic systems—geomorphology should follow up and take advantage of recent (and also older) developments within physics. If the research object is assumed to be assessed best as thermodynamic system, thermodynamic definitions and terms have to be used and cannot be mixed with incompatible assumptions from mechanics. As soon as the system as a whole is focused, the clearly defined equilibria and states of thermodynamics have to be considered and used. The advantage of these definitions is furthermore that they are valid independently of any spatial or temporal scales (also cf. [54]).

However, there was nothing 'wrong'about using the concept of equilibrium in the 1950ies, and that also holds true today, if applied to specific, clearly defined research questions: Especially in the area of landslides, a mechanic approach is suitable as gravitational force is the driving factor that has to exceed certain resisting forces, so that finally the landslide occurs. The important role of gravitation for many geomorphological issues may thus partly explain the focus on mechanic equilibrium as well as the confusion of mechanical and thermodynamic basics. However, it is essential to note that an assessment of acting forces may indeed explain why and how a process is taking place but that *this cannot include any conclusions on the system as a whole*. Within geomorphology there thus must be gained clarity on which level the assessment is taking place: the level of mechanics, thermodynamics, or non-linear thermodynamics. This is not simply pedantry, but a question of scientificality. If one is aware of the level of consideration, this at the same time indicates the scope of the results and conclusions, that is, for the outer limits of results. For example, the assessment of the balance of forces does not allow conclusions on the state the system is currently in—but on the state of single variables. Conclusions on how far a system is from thermodynamic equilibrium are impossible—but it can indeed be concluded whether a mechanic equilibrium between driving and resisting forces exists. However, to claim this as a kind of normal state then becomes simply impossibly due to the awareness of the high level of reduction. Thus, it is paramount to solve this confusion in geomorphology. From this it can be concluded: Only if we gain a strict and consistent terminology, will developments from the 'pure sciences' benefit geomorphology and the connectivity to our main neighbouring disciplines might be increased.

References

1. Prigogine I, Stengers I (1990) Entwicklung und Irreversibilität. In: Niedersen U, Pohlmann L (eds) Selbstorganisation und Determination. Selbstorganisation. Jahrbuch für Komplexität in den Natur-, Sozial- und Geisteswissenschaften. Duncker and Humblot, Berlin, pp 3–18
2. Prigogine I, Stengers I (1993) Das Paradox der Zeit. Zeit, Chaos und Quanten. Piper, München, p 338
3. Gilbert GK (1877) Geology of the Henry Mountains. Government Printing Office, Washington, p 160
4. Kennedy BA (1992) Hutton to Horton: views of sequence, progression and equilibrium in geomorphology. Geomorphology 5(3–5):231–250
5. Bracken LJ, Wainwright J (2006) Geomorphological equilibrium: myth and metaphor? Trans Inst Br Geographer 31:167–178
6. Hack JT (1960) Interpretation of erosional topography in humid temperate regions. Am J Sci, Bradley 258-A: 0–97
7. Ritter DF, Kochel RC, Miller JR (1995) Process geomorphology. Brown, Dubuque, p 546
8. McCoy ED, Shrader-Frechette K (1992) Community Ecology, Scale, and the Instability of the stability concept. In: PSA: Proceedings of the Biennial Meeting of the Philosophy of Science Association, (1):184–199
9. Niedersen U, Pohlmann L (1990) Komplexität, Singularität und Determination. Die Koordination der Heterogenität. In: Niedersen U, Pohlmann L (eds) Selbstorganisation und Determination. Selbstorganisation. Jahrbuch für Komplexität in den Natur-, Sozial- und Geisteswissenschaften. Duncker and Humblot, Berlin, pp 25–54
10. Krug H-J, Kruhl JH (2000) Einführung. In: Krug H-J, Kruhl JH (eds) Nichtgleichgewichtsprozesse und dissipative Strukturen in den Geowissenschaften. Non-Equilibrium Processes and Dissipative Structures in Geoscience. Selbstorganisation. Jahrbuch für Komplexität in den Natur-, Sozial- und Geisteswissenschaften. Duncker and Humblot, Berlin, pp 7–12
11. von Uexküll J (1949) Niegeschaute Welten. Die Umwelten meiner Freunde. Ein Erinnerungsbuch. Suhrkamp, Berlin, p 260 Frankfurt/Main
12. Chin A (2006) Urban transformation of river landscapes in a global context. Geomorphology 79:460–487
13. Graf WL (1977) The rate law in fluvial geomorphology. Am J Sci 277(2):178–191
14. Jantsch E (1979) Die Selbstorganisation des Universums. Vom Urknall zum menschlichen Geist. Hanser, Darmstadt, p 464
15. Gilbert GK (1909) The convexity of hillslopes. J Geol XVII(4):344–350
16. Thorn CE, Welford MR (1994) The Equilibrium Concept in Geomorphology. Ann Assoc Am Geogr 84(4):666–696
17. Davis WM (1899) The geographical cycle. Geog J 14(5):481–504
18. Orme AR (2007) The rise and fall of the Davisian cycle of erosion: prelude, fugue, coda, and sequel. Phys Geogr 28(6):474–506

19. Horton RE (1945) Erosional development of streams and their drainage basins; hydrological approach to quantitative morphology 56:257–370
20. Mackin JH (1948) Concept of the Graded River. Bull Geol Soc Am 59:463–512
21. Hack JT (1975) Dynamic equilibrium and landscape evolution. In: Melhorn WC, Flemal RC (eds) Theories of Landform Development. Allen and Unwin, London, pp 87–102
22. Kesseli JE (1941) The concept of the graded river. J Geol XLIX(6):561–588
23. Strahler AN (1950a) Equilibrium theory of erosional slopes approached by frequency distribution analysis. Part I Am J Sci 248:673–696
24. Strahler AN (1950b) Equilibrium theory of erosional slopes approached by frequency distribution analysis. Part II: significance tests applied to slope problems in the Verdugo and San Rafael Hills, California. Am J Sci 248:800–814
25. Strahler AN (1952) Dynamic basis of geomorphology. Bull Geol Soc Am 63:923–938
26. Ahnert F (1994) Equilibrium, scale and inheritance in geomorphology. Geomorphology 11(2):125–140
27. Bremer H (1984) Das Gleichgewichtskonzept in Zeit und Raum. Zeitschrift für Geomorphologie N.F. Suppl.-Bd. 50:11–18
28. Ahnert F (1954) Zur Frage der rückschreitenden Denudation und des dynamischen Gleichgewichts bei morphologischen Vorgängen. Erdkunde 8:61–64
29. Ahnert F (1970) Functional relationships between denudation, relief, and uplift in large mid-altitude drainage basins. Am J Sci 268:243–263
30. Ahnert F (1984) Local relief and the height limits of mountain ranges. Am J Sci 284:1035–1055
31. Ahnert F (1987) Approaches to dynamic equilibrium in theoretical simulations of slope development. Earth Surf Proc Land 12:3–15
32. Culling WEH (1957) Multicyclic streams and the equilibrium theory of grade. J Geol 65:259–274
33. Schneider ED, Sagan D (2005) Into the cool. Energy flow, thermodynamics, and life. The University of Chicago Press, Chicago, p 362
34. Culling WEH (1987) Equifinality: Modern Approaches to dynamical Systems and Their Potential for Geographical Thought. Trans Inst Br Geographer 12(1):57–72
35. Mayer L (1992) Some comments on equilibrium concepts and geomorphic systems. Geomorphology 5:277–295
36. von Bertalanffy L (1940) Der Organismus als physikalisches System betrachtet. Die Naturwissenschaften 28(33):521–531
37. von Bertalanffy L (1950) The theory of open systems in physics and biology. Science 111(2872):23–29
38. Chorley RJ (1962) Geomorphology and General Systems Theory. Geological Survey professional paper. United States Government Printing Office, Washington, pp 1–10
39. von Bertalanffy L (1954) Das Fließgleichgewicht des Organismus. Kolloid-Zeitschrift 139:86–91
40. Schumm SA, Lichty RW (1965) Time, space, and causality in geomorphology. Am J Sci 263:110–119
41. Howard AD (1965) Geomorphological systems—equilibrium and dynamics. Am J Sci 263(4):302–312
42. Bull WB (1991) Geomorphic responses to climatic change. Oxford University Press, Newyork, Oxford, p 326
43. Howard AD (1982) Equilibrium and time scales in geomorphology: application to sand-bed alluvial streams. Earth Surf Proc Land 7:303–325
44. Chorley RJ, Kennedy BA (1971) Physical geography—A Systems Approach. London
45. Trimble SW (1975) Denudation studies: can we assume stream steady state? Science 188:1207–1208
46. Schumm SA (1979) Geomorphic thresholds. The concept and its applications. Trans Inst Br Geographer 4(4): 85–515

47. Dikau R (2006) Komplexe Systeme in der Geomorphologie. Mitteilungen der Österreichischen Geographischen Gesellschaft 148:125–150
48. von Elverfeldt K, Keiler M (2008) Offene Systeme und ihre Umwelt—Systemperspektiven in der Geomorphologie. In: Heike E, Ratter BMW, Dikau R (eds) Umwelt als System—System als Umwelt? Systemtheorien auf dem Prüfstand. Oekom, München, pp 75–102
49. Phillips JD (2007) Geomorphic equilibrium in southeast Texas rivers. University of Kentucky, Lexington
50. Grams PE, Schmidt JC (2005) Equilibrium or indeterminante? Where sediment budgets fail: sediment mass balance and adjustment of channel form, Green River Downstream From Flaming Gorge Dam, Utah and Colorado. Geomorphology 71:156–181
51. Hooke RL (1968) Steady-state relationships on arid-region alluvial fans in closed basins. Am J Sci 266:609–629
52. Willett SD, Brandon MT (2002) On steady states in mountain belts. Geology 30(2):175–178
53. Young A (1970) Concepts of equilibrium, grade and uniformity as applied to slopes. Geog J 136(4):585–592
54. Montgomery K (1989) Concepts of equilibrium and evolution in geomorphology: the model of branch systems. Prog Phys Geogr 13(1):47–66
55. Phillips JD (1992) The end of equilibrium? Geomorphology 5(3–5):195–201
56. Smithson P, Addison K, Atkinson K (2002) Fundamentals of the physical environment. Routledge, London, p 627

Chapter 7
Fifth Problem Area: Complexity and Non-Linearity

The biggest finder still is a blind man in all the richness of the world

Christian Morgenstern

In the preceding chapter we discussed that those systems which we generally observe in geomorphology cannot be regarded as systems in equilibrium. This is the case, because geomorphic systems are centres of flow, growth, and change—they are neither static, nor still, nor 'dead' (cf. [1, p. xii]). Thus, they are not in equilibrium. Non-linear systems are the norm, not the exception. With increasing results, which contradicted the equilibrium concept, this insight lead to an approach oriented towards non-linearity in the 1990s. When a non-linear approach is applied, every cause can become an effect and every effect can become a cause [2, p. 113], and a system's equilibrium cannot be established. This insight thus was already communicated 40 years ago, but has, however, not been successfully distributed within the prevalent paradigm. Schumm and Lichty [2, p. 112] stressed that

> a choice must be made whether only components of a landscape are to be considered or whether the system is to be considered as a whole.

Until the 1990s, the choice within geomorphology clearly fell on the study of system components with the aim of assessing (mechanical) equilibria. Since then, there have emerged an increasing number of studies on the nonlinear, dynamical behaviour of systems (as a whole) (e.g. cf. [3–9]). However, an oxymoron was created: whilst assuming nonlinearity, these studies, at the same time, do not necessarily deny system states, which can be described as steady or in equilibrium. This contradiction to non-equilibrium thermodynamics therefore also proceeded after the paradigmatic turn, however, these steady or balanced system states are no longer regarded as being "normal" states, whereby to achieve one can 'support' nature.

Hence, the focus shifted from systems in equilibrium to systems far from equilibrium and to systems in non- or disequilibrium (cf. [4]).[1] In this respect, the

[1] It is not clear, however, whether these equilibria are geomorphic or thermodynamic, with the latter being a reception of Ilya Prigogine's approach to dissipative structures. To my knowledge, these differences—between geomorphic and thermodynamic equilibria—have never been assessed.

K. von Elverfeldt, *System Theory in Geomorphology*,
Springer Theses, DOI: 10.1007/978-94-007-2822-6_7,

efforts of the 23rd Binghamton Symposium on "Geomorphic Systems" are of uttermost importance for geomorphology ([4, 10–13], amongst others). These clearly revealed and stressed the limits of equilibrium thinking, especially with regard to an understanding of landscape development. In particular Mayer [11, p. 277] expressed a certain amazement about how equilibrium thinking ever attained such a strong influence on geomorphology as it is highly disproportional to the scientific value of that concept. For many geomorphological research questions the equilibrium concept is simply counterproductive, as it mantles the view on dynamics and change in/of systems. Thus a new approach seems necessary and is increasingly followed within geomorphology (cf. e.g. [14]).

Just as global change is a key for the understanding of earth history, system change can become they key to an understanding of system behaviour. With an accordingly changed approach, geomorphology can go through a development similar to the one which physics already has been through:

> Classical physics has emphasized stability and permanence. We now see that, at best, such a qualification applies only to very limited aspects. Wherever we look, we discover evolutionary processes leading to diversification and increasing complexity [15, p. 1].

Commonly, nonlinearity and complexity are understood as being synonymous, and in that sense mostly as the opposite of something being 'simple'. However, nonlinear systems are not necessarily complex. According to Baas [16], geomorphic complex systems are characterized by three main attributes, which show striking similarities to Prigogine's theory of dissipative structures (cf. Chap. 6): (1) The systems are open and produce entropy, (2) the systems utilize a high number of elements, which are (3) coupled with each other via interactions and feedbacks. Furthermore, complex systems show emergent behaviour, exhibit dissipative structures and are self-organised [16]. Accordingly, there exist three main characteristics for nonlinearity, which are of importance for the observation of geomorphic systems (cf. [17]) and which are also in accordance with Prigogine's non-equilibrium thermodynamics: Firstly, a nonlinear system can exhibit sensitivities to the initial conditions, i.e. small changes of the initial parameters can lead to disproportionally high alternations of the result:

> As accurate as the determination of the initial conditions of such a system might be, we will only be able to infer an indefinite behaviour. The smallest inaccuracy will express itself not only in quantitative, but also in qualitative indeterminacy ([18, p. 9], translation by KvE).

Secondly, the system's behaviour can lead to the emergence of characteristics or structures. Emergent characteristics or structures cannot be explained by the sum of all components [19–21]. Finally, concluding from (1) and (2), large scale and long-term behaviour of systems cannot be predicted by small scale and short-term processes [17]. According to Lane and Richards [17], this last point is the main implication of nonlinearity, which therefore stands against the traditional focus on the prediction of system behaviour or, accordingly, highly diminishes the predictability of systems (cf. also Sect. 10.1).

Furthermore, the perspective of nonlinear and complex systems refers to the reasons why systems behaviour changes, which contrasts the traditional geomorphological system understanding (which, however, can still be found in younger publications, e.g. [9, 20, 22]). Traditionally, these reasons for change are seen in changes or disturbances of the environmental conditions, which control the system. This perspective, however, neglects the consideration that external disturbances often are a mere initiation for general instabilities, but that these externalities are not necessarily the key to an explanation of the system state [23]. This was already emphasized by Schumm [24] thirty years earlier, when he assigned certain changes in system behaviour to the exceedance of specific intrinsic thresholds (cf. Sects. 4.2 and 6.2). According to Schumm [24, p. 478], these thresholds are due to internal developments, i.e. "it is a threshold that is developed within the geomorphic system by changes in the morphology of the landform itself through time".

Thereby the focus is shifted from the environment of systems towards the internal behaviour of the system as well as the resultant forms and structures. The extent to which the environment can determine system behaviour is considerably reduced, just as causalities are constricted—the system acts in a self-organized manner and, in doing so, the system can also reach 'thresholds' at which it changes considerably. In that, Schumm's description of system behaviour is largely congruent with the notion of self-organised criticality (SOC), which was coined eight years later by Per Bak [25, 26]. SOC describes those processes by which a system organizes itself in a critical state (cf. [25]). However, it is neither trivial nor simple to discover self-organisation in geomorphic systems: it is not the case that every recurring pattern and/or structure can necessarily be attributed to self-organisation. Merino and Wang [27], for example, stress that varves in lake sediments cannot be regarded as sign for self-organisation, as they regard these as patterns caused by external conditions, i.e. regular erosion events caused due to climate conditions. However, from my point of view, the problem of system boundaries surfaces here: As long as geomorphologists do not have strict and stringent definitions of "geomorphic systems" and thus a means to delimit system boundaries independently from their respective research question (cf. Chap. 3), geomorphologists will not be able to determine whether systems act in a self-organized manner or rather "are being organized" by external factors. After all, a multitude of studies and observations are necessary for defining a structure or system as self-organised (cf. [27]): sundry involved processes have to be identified and described, feedback-mechanisms discerned, and—if indicated—the system has to be modelled. But there is one prerequisite for each of these steps: knowledge of system boundaries.

It cannot be overemphasized that the contemplation of nonlinearity and complexity, respectively, is a substantial enhancement of geomorphology and marks a rapprochement to the physical basis. Still, geomorphologists remain uncertain in using and applying this physical knowledge. For example, the question why order as well disorder can be observed in geomorphic systems provokes amazement, and several explanations are offered (e.g. [20, p. 58]). However, according to the theory of dissipative structures, order and disorder can exist simultaneously, and order is created from disorder through dissipation. Thus, there is not contradiction at all. Disequilibrium and any associated disorder can be a source of order, and this order is characterized by the occurrence of dissipative structures [28–30].

Despite these tendencies towards the study of nonlinear systems even in contemporary geomorphological research far too often open systems are thought to be in steady state or (in some kind of) equilibrium if the input of mass and/or energy is balanced by self-adjustment or self-regulation of form or geometry.[2] Often, this is due to a crude understanding of the thermodynamic equilibrium or other (rather unspecific) equilibria or steady states (cf. e.g. [31–36]). As has been stated in the previous chapter, however, equilibrium thinking and the theory of open thermodynamic systems are incompatible—it is an internal contradiction of geomorphological system theory, which probably poses one of the major problems in geomorphological theory: Two incompatible theories exist simultaneously and are often *both utilized at the same time* for one and the same study (such as in [32]). If empirics are guided by such a theoretical palimpsest, of which value are the results, especially if perceived by neighbouring disciplines like physics?

Thus, apparently, it is not only the land surface which can be described as palimpsest, but also the theoretical basis of geomorphology (cf. [37]). Originally, the term palimpsest describes a vellum, which is constantly rewritten whilst the older writings are not completely erased. According to Chorley et al. [38] and Dikau [39, 40], the same is true for the land surface: Old forms are not completely "erased" and therefore still influence present forms and processes. Consequently, if assigned to the theoretical basis of geomorphology, a similar picture can be drawn: Due to the fact that few geomorphologists truly engage in (system) theory, neither are outdated theoretical concepts "erased", nor does awareness even exist that these concepts are outdated. Thus, old concepts are overwritten, and as their basic assumptions are not questioned or challenged, terms and assumptions, which are incompatible with newer concepts, still remain.

Instead of merely taking over theoretical concepts from our neighbouring disciplines, geomorphologists should, first of all, engage in their own theoretical basis. Just as defining a "geomorphic system". Just as deciding which knowledge and insights geomorphologists want to attain, which in return is a guide towards the appropriate theoretical basis: Are we aiming at a description of mass and energy flows, or are we aiming at an explanation of the structures of the land surface? If physical basics are applied, geomorphic phenomena such as relatively stable river profiles, flood plains, and alluvial cones can be explained as dissipative structures, which are *not necessarily stable* and which are an *expression of nonlinearity and self-organisation*, not of system's equilibria. Congruently, autopoiesis and self-reference (cf. Chap. 4) can be regarded as specific sets of processes, which create structure in systems far from thermodynamic equilibrium (cf. [41]).

[2] In contrast to self-organisation, self-regulation always stays within the stable band of a system (cf. [18, p. 72]).

The consequences of an adaption of these thoughts for geomorphology will be explained in the following two chapters.

References

1. Schneider ED, Sagan D (2005) Into the cool. Energy flow, thermodynamics, and life. The University of Chicago Press, Chicago, p 362
2. Schumm SA, Lichty RW (1965) Time, space, and causality in geomorphology. Am J Sci 263:110–119 February
3. Hergarten S (2003) Landslides, sandpiles, and self-organized criticality. Nat Hazards Earth Syst Sci 3(3):505–514
4. Phillips JD (1992) The end of equilibrium? Geomorphology 5(3–5):195–201
5. Phillips JD (1999) Divergence, convergence, and self-organization in landscapes. Ann Assoc Am Geogr 89(3):466–488
6. Phillips JD (2006) Deterministic chaos and historical geomorphology: a review and look forward. Geomorphology 76:109–121
7. Phillips JD (2006) Evolutionary geomorphology: thresholds and nonlinearity in landform response to environmental change. Hydrol Earth Syst Sci 10:731–742
8. Schumm SA (1991) To interpret the earth. Ten ways to be wrong. Cambridge University Press, Cambridge
9. Thomas MF (2001) Landscape sensitivity in time and space—an introduction. Catena 42(2–4):83–98
10. Malanson GP, Butler DR, Georgakakos KP (1992) Nonequilibrium geomorphic processes and deterministic chaos. Geomorphology 5:311–322
11. Mayer L (1992) Some comments on equilibrium concepts and geomorphic systems. Geomorphology 5:277–295
12. Renwick WH (1992) Equilibrium, disequilibrium, and nonequilibrium landforms in the landscape. Geomorphology 5:265–276
13. Sack D (1992): New wine in old bottles: the historiography of a paradigm change. Geomorphology, 5:251–263
14. Phillips JD (2009) Changes, perturbations, and responses in geomorphic systems. Prog Phys Geogr 33(1):17–30
15. Nicolis G, Prigogine I (1977) Self-organization in nonequilibrium systems. From dissipative structures to order through fluctuations. Wiley, New York, p 491
16. Baas ACW (2007) Chaos, fractals and self-organization in coastal geomorphology: simulating dune landscapes in vegetated environments. Geomorphology 48(1–3):309–328
17. Lane SN, Richards KS (1997) Linking river channel form and process: time, space and causality revisited. Earth Surf Proc Land 22(3):249–260
18. Prigogine I, Stengers I (1990) Entwicklung und Irreversibilität. In: Niedersen U, Pohlmann L (ed) Selbstorganisation und Determination. Selbstorganisation. Jahrbuch für Komplexität in den Natur-, Sozial- und Geisteswissenschaften. Duncker and Humblot, Berlin, pp 3–18
19. Harrison S (2001) On reductionism and emergence in geomorphology. Trans Inst British Geographers 26(3):327–339
20. Phillips JD (1999) Earth surface systems: complexity, order and scale. Blackwell, Oxford, p 180
21. Phillips JD (2003) Sources of nonlinearity and complexity in geomorphic systems. Prog Phys Geogr 27(1):1–23
22. Dikau R (2006) Komplexe Systeme in der Geomorphologie. Mitteilungen der Österreichischen Geographischen Gesellschaft 148:125–150

23. Murray B, Fonstad MA (2007) Preface: complexity (and simplicity) in landscapes. Geomorphology 91:173–177
24. Schumm SA (1979) Geomorphic thresholds. The concept and its applications. Trans Inst British Geographers 4(4):485–515
25. Bak P, Tang C, Wiesenfeld K (1987) Self-organized criticality: an explanation of 1/f noise. Phys Rev Lett 59:381–384
26. Bak P (1996) How nature works. The science of self-organised criticality. Copernicus Press, New York, p 212
27. Merino E, Wang Y (2000) Geochemical self-organization in rocks: Occurrences, oberservations, modeling, testing–with emphasis on Agate genesis. In: Krug H-J, Kruhl J H (ed), Nichtgleichgewichtsprozesse und dissipative Strukturen in den Geowissenschaften. Non-equilibrium processes and dissipative structures in geoscience. Selbstorganisation. Jahrbuch für Komplexität in den Natur-, Sozial- und Geisteswissenschaften. Duncker and Humblot, Berlin, pp 13–46
28. Prigogine I (1967) Introduction to thermodynamics of irreversible processes. Interscience, New York, p 147
29. Prigogine I (1985) Vom Sein zum Werden. Zeit und Komplexität in den Naturwissenschaften. Piper, München, p 304
30. Prigogine I, Stengers I (1980) Einleitung: Die Herausforderung an die Wissenschaft. In: Prigogine I, Stengers I (ed) Dialog mit der Natur, München, pp 9–30
31. Chin A (2006) Urban transformation of river landscapes in a global context. Geomorphology 79:460–487
32. Dearing JA, Zolitschka B (1999) System dynamics and environmental change: an exploratory study of Holocene lake sediments at Holzmaar, Germany. The Holocene 9(5):531–540
33. Grams PE, Schmidt JC (2005) Equilibrium or indeterminante? Where sediment budgets fail: sediment mass balance and adjustment of channel form, Green River downstream from flaming Gorge Dam, Utah and Colorado. Geomorphology 71:156–181
34. Petts GE, Gurnell AM (2005) Dams and geomorphology: research progress and future directions. Geomorphology 71:27–47
35. Tucker GE (2009) ESEX commentary: natural experiments in landscape evolution. Earth Surf Proc Land 34:1450–1460
36. Wang Z-Y, Wu B, Wang G (2007) Fluvial processes and morphological response in the Yellow and Weihe Rivers to closure and operation of Sanmenxia Dam. Geomorphology 91:65–79
37. von Elverfeldt K, Keiler M (2008) Offene Systeme und ihre Umwelt–Systemperspektiven in der Geomorphologie. In: Egner H, Ratter BMW, Dikau R (eds) Umwelt als System–System als Umwelt? Systemtheorien auf dem Prüfstand. Oekom, München, pp 75–102
38. Chorley RJ, Schumm SA, Sudgen DE (1984) Geomorphology. London, New York, p 605
39. Dikau R (1996) Geomorphologische Reliefklassifikation und -analyse. Heidelberger Geographische Arbeiten, p 104
40. Dikau R (2005) Geomorphologische Perspektiven integrativer Forschungsansätze in Physischer Geographie und Humangeographie. In: Wardenga U, Müller-Mahn D (ed) Möglichkeiten und Grenzen integrativer Forschungsansätze in Physischer Geographie und Humangeographie. Forum ifl. Leibniz-Institut für Länderkunde, Leipzig, pp 91–108
41. Boulding KE (1980) Foreword. In: Zeleny M (ed) Autopoiesis. Dissipative structures and spontaneous social orders. Westview, Boulder, pp 17–21

Chapter 8
Tentative Conclusions in Two Steps

8.1 More Common Ground Than Divisiveness: Comparing Second Order System Theories in Physics and Biology

The newer system theories from physics, biology, and also from sociology can be summarized under the generic term "second order system theories", as all of them have been developed within the paradigm of self-organisation. Still, they are different: They start from different basic assumptions and, most of all, are concerned with completely different research objects. For example, one focuses on biological systems, whilst the others deal with thermodynamic or social systems. Importantly, this does not imply that second order system theories are contradictory. Consequently, if geomorphology adopted some thoughts and approaches from biological *and* physical system theory, no new logical inconsistencies or contradictions should arise. If biological second order system theory is compared to the theory of dissipative structures it comes as a surprise that, despite the completely different approaches and chosen routes of theory development, there is much common ground.

As I presented in Sects. 4.2 and 4.3, the theory of dissipative structures belongs to non-equilibrium thermodynamics, or rather non-linear thermodynamics. Thermodynamics, by definition, focus on systems, thus the theory of dissipative structures can be assigned to the multitude of system theories. The focus of this theory is (1) dissipation, that is, the transformation of exergy (energy capable of working) to energy incapable of working, (2) self-organisation, as well as (3) resulting structures and (4) the evolution of open systems. With hindsight, the basic assumption of Prigogine seems to be rather simple, but it revolutionized not only parts of physics: The idea to transfer the second law of thermodynamics from isolated systems to open systems. In doing so, internal entropy production is focussed—thus, in the widest sense energy consumption—but also the entropy exchange across the system borders.

The biological and sociological theories based on the concepts of autopoiesis and self-reference are also system theories. However, they fundamentally differ

K. von Elverfeldt, *System Theory in Geomorphology*,
Springer Theses, DOI: 10.1007/978-94-007-2822-6_8,

from first order system theories that rely on the unity of elements and their interrelations. Biological and sociological system theories do not start from a unity (of elements) delimited by an observer, but are characterized by a difference, namely the difference between the system itself and its environment. This implies a demarcation, whereby space is structured in a different way than before. However, these boundaries are drawn by the system itself and not by an observer in a more or less arbitrary way, depending on the respective research aim or question. But this does not imply that systems are assumed to exist in reality. Quite the contrary: Social and biological second order system theories emphasize that systems or the concept of systems simply is a (more or less) useful tool for the description of the world and that it should not be inferred that systems do really exist.

First difference: Ontology and epistemology
Whilst the theory of dissipative structures does not include explicit epistemological and ontological statements, those theories assuming autopoiesis and self-reference are much clearer on this issue.

First area of common ground: Order through system formation
Despite this fundamental difference another area of common ground solidifies: The assumption that order is built as soon as a system develops and forms. In a more teleological wording, one could say that systems form to build order. Prigogine understands order as the building of structures, whereas Luhmann suggests a reduced complexity as compared to the environment. Furthermore, Maturana and Luhmann suggest the respective process that builds order and structure: autopoiesis.

Second area of common ground: Self-organisation
Prigogine was able to demonstrate that systems far from equilibrium may show fluctuations that subsequently form spatial structures. I presented the case of a hurricane with its characteristic structures as an example. Simplified, the inner entropy production rises with the steepness of the gradients—and in consequence the system becomes more effective at reducing the initial differences. In order to express the inherent contradiction of order due to entropy production, Prigogine calls these structures dissipative. With these structures the system organises itself as it enhances initial fluctuations and establishes its own internal rules about how to deal with itself and with its environment. The price to be paid is an increased entropy production, and in consequence this entropy is transferred to the environment. Again, this can be illustrated with the example of a hurricane that literally creates a lot of disorder on his way. Prigogine was able to show that beyond the linear branch of thermodynamics no physical law exists that determines a specific system behaviour. Self-organised systems exhibit highly specific behaviour and there is no rule from which a system could be an exception. The specific system behaviour depends on the structure and the history of the system, and in some phases of system evolution it is simply coincidental. The framework conditions, namely the environment, cannot determine system behaviour: There are

several solutions (i.e. system states) for the same parameter values. Self-organised systems are structurally determined and thus autonomous.

Within biological second order system theories systems are also regarded as self-organised: The individual system gives itself a specific and characteristic structure. Again, this structure is not static but may (but does not have to) change according to the strategies of the system to deal with irritations.

Thus, the aspect of self-organisation offers a high potential for connecting the different approaches.

Second difference: Self-reference

However, the approaches of autopoieses even go one step further, since they assign systems the capability of self-reference in all their operations. After all, this creates the boundary between system and environment, the system becomes a unity. At the same time, due to this strict boundary, the system is autonomous from its environment, and to a much larger extent than a self-organised system. It is plausible that a system that only refers to itself in all its actions is much harder to irritate than a self-organised system. Within a self-organised system, the consequences of certain actions might not be predicted (also cf. Sect. 9.1), but still: interventions are not precluded per se by the way it functions. Self-reference implies that a system is setting itself according to its structure whether it reacts to external signals or not. The specific reaction or non-reaction is, in some sense, nothing but an update of one of several different possibilities. In consequence, autopoiesis and self-reference imply an operative closeness of the systems since every system type has its own mode of operation, and this mode of operation is absolutely exclusive: a different type of system necessarily has a different mode of operation. That is, cells cannot think or communicate. But from this it also becomes clear that operative closeness is indeed restricted to the operations of the system and does not include or mean that it is closed for matter and energy flows.

Third area of common ground: Historicity of systems

Due to the existence of structures as well as their non-linear properties, autopoietic systems can be assumed to be far from thermodynamic equilibrium and to be separated from their environment by energy gradients (cf. [1]). Furthermore, they can proceed spontaneously from one ordered state to another [2]. Additionally, autopoietic systems are historical, as the series of states are indeed related to interactions with the environment, but are neither determined by these interactions nor reversible, and they could principally have been totally different ([3, p. 71], according to Bühl [1]).

Third difference: determinacy of systems

From the perspective of biological second order system theory, systems are indeed permeable for mass and energy flows, but this is still basically different from Prigogine's concept. Autopoietic and/or self-referential systems can neither be determined by these flows, nor by the framework conditions as set by the environment. The theory of dissipative systems, in contrast, does assign phases within the evolution of systems during which they can be completely described by the framework conditions. Within the concept of autopoietic systems, mass and energy are externalised, that is, they are a necessary, but not a sufficient condition for the existence of a system and for the respective system state.

Fourth area of common ground: Autonomous system behaviour
It is true for both approaches that the self-imposed and self-created structure determines the reactions of the system[1]: Which and if at all processes or operations are possible, depends on the structure of the system. Thus, the focus is directed towards the inner organisation, towards the processes and the evolution of the system.

Fifth area of common ground: unpredictability of the evolution
Self-organised thermodynamic systems behave very specifically, because there is no universal physical law beyond linearity that describes the evolution of systems and allows for a calculation of state. Most of all, self-organised thermodynamic systems are not automatically stable anymore, as is the case close to or at equilibrium. To make a point, death is extremely stable—whilst life is subject to continuous changes and thus also allows instabilities. The keyword for systems far from equilibrium thus is change, including changes towards instable states. The reasons for instability are manifold: it can organize itself to instability (e.g. self-organised criticality), it can be driven towards instability by changed environmental conditions, or it reacts to new, alien material. Beyond certain instability thresholds, the system finally reaches so-called bifurcation points where the system 'chooses' its state, even chooses its future. The decision for one of the possible new states is arrived at by chance and thus is not predictable. At the same time, simply *because* the system chooses just one of several states it has a kind of memory: This one choice, this one event influences the whole future evolution, a characteristic called path dependency. Thus, the history of the system becomes crucial for the understanding of current and future system states, because each bifurcation leads to completely different possibilities for the system. In anticipation of Sect. 10.1, which deals with predictability, it can be concluded that the current as well as every future state of a self-organised, dissipative system is a result of chance and (mostly not known) history.

The characteristic of unpredictability is also true for self-referential, autopoietic systems—it cannot be predicted which path a system chooses based on which environmental information.

In comparison it becomes apparent that the two theories partly use different terms, but that they in no case are mutually exclusive and often even come to

[1] Strictly speaking, determination means complete determination of the future state (also cf. Sect. 8.3, where the term is explained in connection with causality). Maturana as well as probably Prigogine, however, rather use the term 'structurally determined' in a rhetorical manner in order to stress and illustrate the importance of different influencing factors (also cf. [4]).

similar conclusions. A conflation of the approaches seems to be possible, or maybe just the development of geomorphological system theory towards a system theoretical approach under the umbrella of the paradigm of self-organisation seems to be feasible. Terms and concepts of both physical as well as biological second order system theories can be principally utilized.

8.2 The Distinction Makes the Difference: Theoretical Inconsistencies of Geomorphological System Theory

As described in Chap. 2, observation theory allows for focussing on the underlying distinctions of the respective theoretical approaches. In the case of this study, thus the main distinctions of the current geomorphological systems theory on the one hand, and the second order system theories on the other hand can be observed. Furthermore, this procedure may emphasize the importance of observation theory for the scientific progress of geomorphology.

So far it has been shown that physical geography can benefit significantly from system theories, however, this potential has not yet been fully tapped. One reason for this is that our discipline widely utilizes reductionist and mechanistic principles. For any scientific theory, the initial assumptions are of uppermost importance, as they are determining the potentials and boundaries of analysis. This is also true for system theories. Within the different approaches, systems are not defined the same, and thus the analytical potential of the respective theoretical approach also varies. To use an example of the biologist Francisco Varela (cf. Varela in [5, p. 149f]): There is nothing to be said against describing a system in a logical and stringent manner by means of inputs, throughputs, and outputs. Between the inputs and the outputs, however, there is a *black* or a *grey box*, and the aim of the system analysis is to reach a *white box* in the end. Subsequently, relations between inputs and outputs can be inferred, or rather a reaction scheme to certain causes can be identified. In this way, however, neither the dynamics of a system nor its self-regulation and self-organisation can be understood. If this is to be considered, a different theoretical approach has to be chosen, which allows for focusing the interior of the system, as well as its autonomy. Thereby, completely new questions arise and, as the case may be, new research methods.

From the perspective of observation theory, it could be principally argued that both points of views are based on the same distinction. They follow the question how a system behaves and develops in a given surrounding. Here, either the environment and its influences on the system can be focused, or the system and its inner organisation and structure. The first option corresponds with the established geomorphological approach and assumes an interrelation of everything, as system boundaries do not have a specific meaning. The second option corresponds with the approach of Maturana as well as Prigogine. It demands a definite demarcation of the system, as otherwise it cannot be decided, which phenomena are a result of self-organisation and which are not. As was shown before, exactly this problem is at the core of geomorphological system theory: The system concept is that inconsistent and arbitrary that it somehow evades precise definition. Consequently, a system cannot be distinguished from its environment.

Based on this assumption, I record a first essential theoretical challenge to geomorphology: geomorphology has to find an answer to the question of system boundaries. For this, second order system theories seem to be helpful, as their starting point is the difference between system and environment, and not the unity of the system. In this theoretical framework, it is the system itself, which draws the line, and the system cannot be thought without its negative correlate, i.e. the environment, and vice versa. The systems 'are there', but they do not add themselves to a single large system, that is, the 'system earth' or the 'environmental system' (for this thought, cf. [6, p. 22]). However, in theory, several systems can take part in the production of a bigger one. For example, a river system or a glacial system can take part in a hydrological system that, nonetheless, has its own laws, and which has to be observed according this laws and characteristics [6, p. 23].

Strahler assigned far-reaching significance to system theory within geomorphology. He pinned his hopes for a success of geomorphology as a discipline to its system theoretical foundation:

> Geomorphology will achieve its fullest development only when the forms and processes are related in terms of dynamic systems and the transformations of mass and energy are considered as functions of time [7, p. 934f].

The reason for this estimation was the aim, or rather the perceived advantage for geomorphology: System theory would lead the focus away from single elements, which seemed to be of importance for the evolution, towards the system and/or the landscape as a whole [8, p. B9]. Thereby, the processes are also focussed, even more than it is de facto currently the case. Although wide parts of geomorphology claim that they conduct process geomorphology, their theoretical foundations with its mechanistic concepts and definitions do not allow for this. The presented concepts from physics and biology with their explicit focus on the modus operandi that defines the system, in contrast, offer a particularly stringent theoretical foundation for the process-oriented research. The mutual equivalence of structure and function, that is, that the evolving structure corresponds exactly to its function and vice versa, is nothing but an expression of this process thinking [9, p. 75]: the structures are a result of interacting processes, which, therefore, only turn the system into what it is. Hence, in this case, the missing distinction makes the difference: Within geomorphology, between mechanistic and thermodynamic concepts is not consequently distinguished. As a result, statements may not be congruent or compatible with the initial assumptions. The demand of the mathematician George Spencer-Brown [10] "Draw a distinction" aims at the insight that only by a distinction something comes into being, or to put it the other way round: if nothing is distinguished, nothing becomes existent. Seen in this light, a new universe can indeed come into being for geomorphology, as soon as the difference between mechanics and thermodynamics is actively and consequently taken into account.

The missing distinction between the different levels of consideration—mechanics on the one hand and thermodynamics on the other hand—can also explain the persistence of the equilibrium concept within geomorphology. Due to this focus on equilibrium states, yet another distinction cannot be perceived: the difference between the stability of structures and the stability of systems. Without the perspective of non-equilibrium thermodynamics, dissipative structures such as dunes, river meanders etc. only represent variables that characterize an equilibrium of any kind. The structure is understood as an expression of the system state, and therefore rather easily measurable variables such as slope or river profiles seemingly can be regarded as state variables—it is assumed that they describe the state of the system. In contrast, in thermodynamics state variables are defined as average values that are independent of the path on which the system reached its state. The mathematical expression for this is a contour integral that becomes zero for state variables. As far as I understand, geomorphological 'state variables' have never been tested in this respect. From this thus results another unclear or missing distinction: that of state variable versus variable.

Considering these inherent problems of geomorphological system theory, Hard [11, p. 95] rightly termed (geomorphic) systems as constructs that solely depend on the respective interests. Therefore, he continues, (geomorphic) systems are not systems in the narrower sense of modern system theories, as there systems are defined by (autopoietic) self-demarcation with respect to its environment. This is why Hard, already as early as in the 1970s, claimed a specification of the geomorphological system theory, as otherwise it remained at best an orientation theory with poor information content. Geomorphological system theory is half-baked and immature, and this results in a juxtaposition and a hodgepodge of inconsistent theoretical assumptions and statements.

With respect to the understanding of complexity there also results an important difference between geomorphological system thinking and second order system theories. Within geomorphology, complexity is mostly used along the difference complex/simple: If something is not complex, then it is simple or linear. Therewith, geomorphological system theory has an 'exculpation capacity': If problems cannot be solved, this is due to the complex structure of the world and the confusing and complex interrelatedness of everything with everything else (cf. [6]). At the same time, this is used as an explanation (and maybe justification?) why to adhere to reductionism. Complexity, in this sense, is the same as 'being complicated'.

Second order system theories open up another view on complexity. On the one hand, system formation already means a reduction of complexity, since the system is always less complex than its environment (cf. [12]). Thus, there exists something like a complexity gradient between the system and its environment, and in this sense the system is a stabilisation of this gradient, the stabilisation of a fundamental asymmetry (cf. [6]). But as systems are also complex, according to Fuchs the distinction is not complex/simple, but 'selective relationability of elements'/'complete relationability of elements'.

Before trying to outline a geomorphological system theory, I want to expatiate the epistemological and practical implications of a change of system perspective. After all, it is exactly on this most fundamental level of scientific knowledge where the changes in the way of thinking become most apparent.

References

1. Bühl WL (1987) Grenzen der Autopoiesis. Kölner Zeitschrift für Soziologie und Sozialpsychologie 39:225–254
2. Roth G (1986) Selbstorganisation–Selbsterhaltung–Selbstreferentialität. In:Dress A, Henrichs H, Küppers G (ed.) Selbstorganisation. Die Entstehung vonOrdnung in Natur und Gesellschaft. Piper, München, Zürich p 149–180
3. Maturana HR (1980) Autopoiesis: reproduction, heredity and evolution. In: Zeleny M (ed) Autopoiesis, dissipative structures and spontaneous social orders, Boulder, pp 45–79
4. Allefeld C (1999) Erkenntnistheoretische Konsequenzen der Systemtheorie. Die Theorie selbstreferentieller Systeme und der Konstruktivismus. Master Thesis, Freie Universität Berlin, Berlin, p 85
5. Simon FB (1997) Autopoiese, strukturelle Kopplung und Therapie–Fragen an Francisco Varela. In: Simon FB (ed) Lebende Systeme: Wirklichkeitskonstruktionen in der systemischen Therapie. Suhrkamp, Frankfurt, pp 148–164
6. Fuchs P (1992) Niklas Luhmann–beobachtet. Eine Einführung in die Systemtheorie. Westdeutscher Verlag, Opladen, p 219
7. Strahler AN (1952) Dynamic basis of geomorphology. Bull Geol Soc Am 63:923–938
8. Chorley RJ (1962) Geomorphology and general systems theory. Geological Survey professional paper. United States Government Printing Office, Washington, pp 1–10
9. Jantsch E (1979) Die Selbstorganisation des Universums. Vom Urknall zum menschlichen Geist. Hanser Verlag, Darmstadt, p 464
10. Spencer-Brown G (1997) Laws of form. Gesetze der Form. Bohmeier Verlag, Lübeck, p 200
11. Hard G (1973) Zur Methodologie und Zukunft der Physischen Geographien an Hochschule und Schule. Möglichkeiten physisch-geographischer Forschungsperspektiven. Geographische Zeitschrift (61): 5–35 (hier aus: Hard G (2003) Dimensionen geographischen Denkens. Aufsätze zur Theorie der Geograpie, Bd. 2, Osnabrück, S. 87–111)
12. Egner H (2008) Komplexität. Zwischen Emergenz und Reduktion. In: Egner H, Ratter BMW, Dikau R (eds) Umwelt als System–System als Umwelt? Systemtheorien auf dem Prüfstand. oekom, München, pp 39–54

Part II
Consequences and Implications

Chapter 9
Epistemological Consequences

There are more things in heaven and earth, Horatio, than are dreamt of in your philosophy!

Shakespeare's Hamlet

In Chap. 2 I presented a concept of observation with far-reaching epistemological implications. These implications then re-appeared within the subsequent chapters. Probably, the most essential consequence of a concept of observation that understands observation as the twofold practice of distinction and simultaneous indication of the afore-distinguished side, is the abandonment of an univocal reality concept. After all, what is actually being observed depends on the distinction: A certain reality is updated, and exactly that one, which comes into being due to this distinction and not others [1], p. 39]. For example, a slope can be observed with the distinction landslide/no landslide, or with loess-covered/not loess-covered, bedrock/no bedrock etc. In this sense, reality is something that comes into being by means of a distinction. Or to put it another way: 'The' reality 'itself' cannot by doubtlessly accessed—which is an extremely inconvenient thought. Furthermore, it contains a tautology: What is not observed is not being observed, and what is observed is being observed [1], p. 40]. This 'inconvenient thought' is the subject of the following chapter.

9.1 On Reality, Objectivity, and Truth

The reality is not inside the things.

Bernard d'Espagnat

In order to detail the epistemological consequences of the afore presented second order system theories I need to start with some thoughts on science and its approaches. In doing so, the epistemological aspects will be focused, that is, the question of how we realise that we perceive. What is the relation between our 'outside', i.e. the surrounding world, and that which is pictured inside our heads and theories?

K. von Elverfeldt, *System Theory in Geomorphology*,
Springer Theses, DOI: 10.1007/978-94-007-2822-6_9,

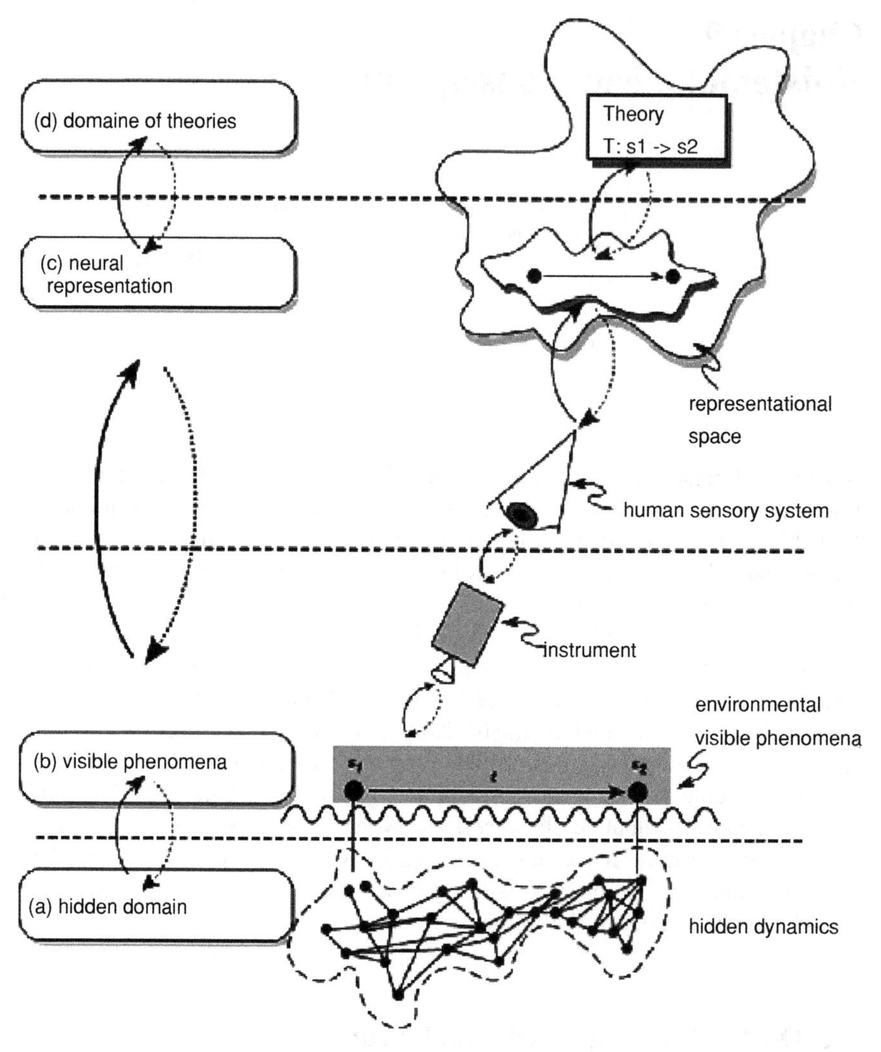

Fig. 9.1 The basic situation of science when observing the world. From the hidden domain to the domain of theories [2], modified

Peschl [2] suggests the distinction between four different domains (cf. Fig. 9.1):

1. The hidden domain **H**: This domain is only accessible for human beings through the use of instruments and (mostly) not directly by our senses (cf. Chap. 2). In the end, this domain becomes the explanandum. The existence of this domain can only be assumed, since we are often confronted with phenomena that cannot sufficiently be explained through themselves.
2. The domain of visible phenomena **V**: This domain is not restricted to visible phenomena but includes all perceivable phenomena. These phenomena are emergent effects of the underlying hidden domain H.

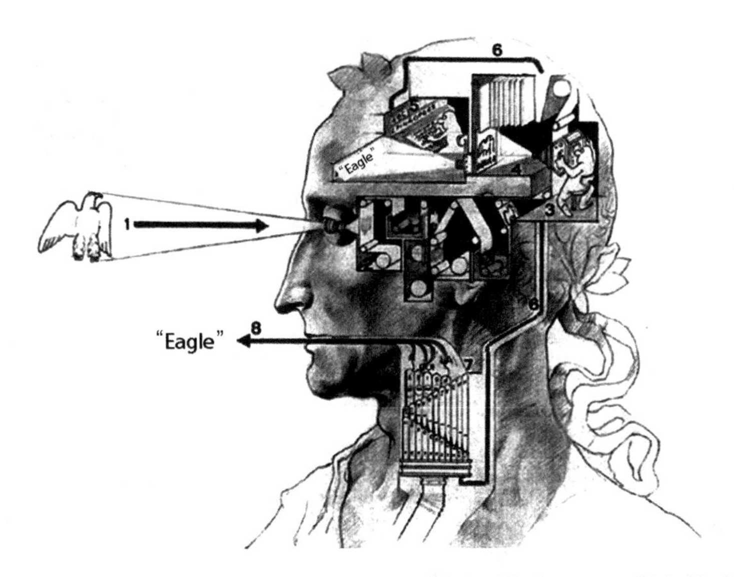

Fig. 9.2 The depiction of reality in our head, exemplified with Caesar watching his favourite symbol, an eagle [5], Varela [6], modified

3. The domain of primary representations **R**: Here, the effects of the hidden domain are represented within the cognitive system.
4. The domain of theories **T** (*theories*): This is the realm of science and the related representation of phenomena by theories (the explanans). The representation can either be internal ('within the head') and/or external, that is, in books, papers, illustrations etc.

The realistic epistemological position assumes that reality (seen as independent from our consciousness) is directly accessible and perceivable (cf. [3]). In a world that is the same for all those things are most esteemed that can be directly seen, heard, and learned [4], p. 5f, citing Plato) (Fig. 9.2).

These phenomena that are either directly accessible or made perceivable by the use of instruments are precisely represented, structured and sorted in our brain (more detailed in e.g. [6], cf. Fig. 9.3). Thus, objective and impartial knowledge of the world, and why the world is as it is, becomes possible. It is only the subject-object-dichotomy that parts us from reality [7]. By the utilization of increasingly better and elaborated observing and measuring systems, we are capable of collecting knowledge, finally approaching omniscience.

The main argument that this reasoning is correct and coherent is that mankind would not have been able to survive (especially not this successfully!) if the representation was not 'correct': in order to survive we have to have an impression of what is going on around us. The question of how this correct representation is coherent with theory-dependent, and in that sense 'biased' perception (also cf. Fig. 9.4), however, remains unanswered (Fig. 9.3).

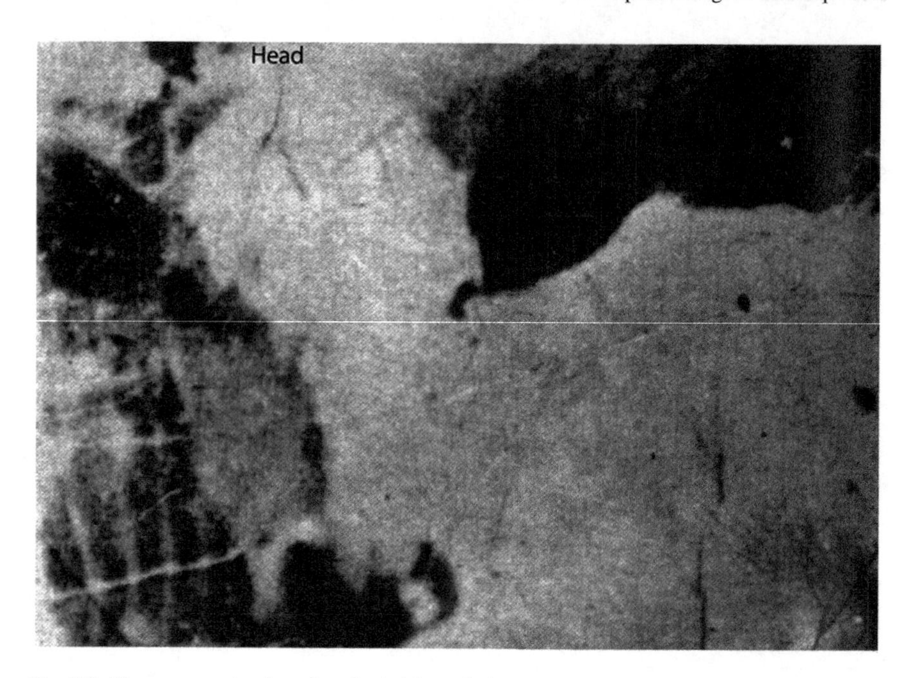

Fig. 9.3 Do you see what I see? ... Probably, only in the very moment in which you know what to see in this picture [8] you will be capable of seeing it. At the same time, as soon as you know what to see, most probably you will not be capable of seeing anything else afterwards (I see a cow—and you? If you cannot see the cow even after being informed that 'it is' a cow you might need someone *to point it out* to you. This, then, corresponds to the process of 'training of perception' as described by Ute Wardenga) (cf. Chap. 2 and [9])

Furthermore, it is neglected that perception (of the domains) and theory are mutually dependent (as illustrated in Fig. 9.2) (cf. [2]). Furthermore, from a theoretical point of view, it is sufficient to merely have an approximate idea of what is going on 'in reality' in order to survive (cf. [10]). The biologist Rupert Riedl illustrates this aspect with the example of our conception of time and space: For us, time seems to be flowing. It comes from somewhere, never comes back and disappears to somewhere else [11, p. 74]. We are quite at a loss when asked where time starts and where it ends. Quite the contrary is the case where space is concerned: We 'see' three-dimensionally, euclidically. Still, the question of where the boundaries of space are, e.g. those of the universe, cannot be answered.

These thoughts merely hint that our views and conceptions only insufficiently match reality—hints that condense with Einstein: According to Einstein, space and time have to be thought of as space–time continuum,[1] which is something that exceeds our imaginative power—and our imagining capabilities—by far. Riedl [11] understands these examples as a warning that our perception can only be a rough

[1] If space is thought of this way, we would be able to see the back of our head if we could only see far enough.

approximation of reality. Our perception is just sufficiently complex in order to survive. Statements on whether we do perceive everything that exists cannot be made. Russell [12, p. 102] puts it this way: The fact that the things we perceive have in common that they are perceived by us is simply a truism from which nothing important can be inferred. Obviously, it is fallacy to infer from the fact that everything we perceive is a perception that therefore everything is perceptible.[2]

As Spencer-Brown, von Foerster, Maturana and Varela explain, a completely different approach to cognition is possible. Von Foerster, Maturana and Varela conducted experiments according to which our perception is not a reflection of environmental features, but a reflection of our own anatomical and functional organisation of the neural system (cf. [14–16]).[3] In order to underpin this point of view, Varela [6] refers to findings on colour vision. If perception is understood as a representation of reality, colour is a characteristic of an object: for example, a green object emits light of the respective wave length. However, Varela [6, p. 59] can show that the relation between wave length and colour does not necessarily exist. For example, in some situations we do perceive 'green' whilst the wave length—according to measurements—is that of the white spectrum (the so-called experiment of coloured shadows). This is one example of why cognition is thought of as all those descriptions a single individual can make (cf. [16]). Cognition, or rather the neuronal representation of the outside (cf. Fig. 9.2) thus becomes specific and subjective as it is determined by the inner structure (e.g. experiences) of any single person, and not by environmental conditions [17]. Thus, the domain of visible phenomena 'disappears' as what we 'see' is nothing other than an internal (sensoric and neuronal) construction:

> You point at a thing, the child looks in the direction you indicate, and the child sees the thing—because the thing is there, just as you see it. I would like to suggest that this is a mistaken oversimplification. What the adult sees and what the child sees are not at all the same thing. In fact, it could not be the same thing, because for the adult the conceptions of things are shaped by a variety of experiences the child has not yet had. In principle, this is similar in the case of two adults, because one person's experiences are never the same as another's. Thus, when you are told that a particular word means "that thing over there", the word's meaning, for you, becomes what you see—and what you see is not what the other sees. What you see is what you have learnt to isolate in your own visual field, by handling things, pushing things, avoiding things—in short, by interacting with your own experiential world, not with anyone else's. And although, as a child, you may have learned the ways you handle, push, and avoid things, to a large extent by copying what you think the adults do among whom you are living, this, too, is a subjective enterprise ([18, p. 23], also cf. [2]).

---

[2] In the following, Russell ([13], p. 123) states that the two possible answers to the question whether real objects exist or not are due to the utilisation of different (causal-theoretical/logical) methods. Therefore, there cannot be an answer to this question. This does not lead physics ad adsurdum: After all, the objects of (or within) our perception form a part of the actual physical material, in fact: the only part that can be empirically encountered. Thus, they deserve the description 'physical'. Still, Bertrand Russell cannot be described as advocate or supporter of (radical) constructivism (or rather, at his time: idealism). Rather, his conceptions are intermediate to realistic and idealistic positions, both of which he regards as mistakes (cf. Russell 13/12).

[3] Paradoxically, these insights have to be applied to themselves.

Maturana [19] argues that the nervous system is a closed system. In this sense, reality can only be that part which is defined by the operations of the observer [20]. Hence, any knowledge is relative and depends on the individual. At the same time, due to autopoiesis and structural coupling this cannot be arbitrary: Only successful actions are repeated [19]. This instantly leads to the question of objectivity (that is, after all, central to (realistic) scientific approaches): What about objectivity, if any scientific statement is inevitably subjective? According to Maturana ([17], also cf. Schmidt [16]), scientific objectivity is a result of the cultural uniformity of the observers, and objectivity in sensu *stricto* does not exist. This thus brings us full circle to the first and especially the second chapter where the theory-dependency of all observations has already been pointed out: Theory-dependency finds its biological and empirical explanation with Maturana [17, 20, 19] (also cf. [16]). Thus, from Maturana's intention to biologically understand cognition (cf. [19, 21, 22]) emerged a biologically founded epistemological theory. Despite completely different pathways, Maturana's theory complies to a large degree with the findings of Spencer-Brown ([23, 24]), von Glasersfeld [25] and von Foerster ([26–28, 14]) (cf. Chap. 2). Common to all these approaches is the insight that it can never be judged whether someone has a completely true image of reality, because its given rightness cannot be verified: After all, no-one can escape his or her own structures of perception and concepts. Any attempt of comparing the image of reality with reality itself is coined by our own perceptive instruments [29, p. 48].

The negation of a representation of reality does not imply, however, a negation of "reality itself", which would indeed be foolish [25, p. 43]. It is impossible that the world does not exist [7]. Consequently, our perception might just as well fit reality, but if this is to be stated it would have to be proved as well. And this proof is impossible. Reality is exactly what we see—or might as well be something completely different. What remains certain is that our perception is sufficient for survival in this reality (at least until now).

This is as much as we can say. This is especially as much as we can say if it has to be judged whether something is right or wrong, as nobody has a 'special access to reality' ([30, p. 51) and nobody can exceed his or her own horizon of experience (cf. [7]). Scientific claims on truth thus have to be abandoned. Von Glasersfeld [31, p. 29] suggests to assume the functioning of ideas' (also termed viability or functional fit) instead of truth. Within the field of experience viability replaces the traditional philosophical concept of truth, which states a 'correct' representation of reality [25, p. 43].

Glanville ([32], also cf. [7]) shows that this attitude does not imply the end of science. Quite the contrary: It can enrich science and contribute to the solution of contemporary fundamental scientific problems.

To conclude, eight statements on the possibilities and boundaries of science and knowledge can be summarized (cf. [2]):

1. Any knowledge is hypothetical;
2. knowledge is subject to continuous change (and this change does not necessarily lead towards 'more knowledge');

3. knowledge only characterizes reality in terms what it is *not* (for example, the Earth is *not* hostile to life);
4. knowledge results from active involvement (as we construct knowledge by acting and interacting with our environment);
5. knowledge is always system-relative (that is, subjective);
6. our knowledge fits its task: our survival;
7. thus, knowledge is a strategy to successfully handle our environmental conditions;
8. two or more equally viable and potentially contradicting theories might exist.

Since our knowledge is inevitably reduced due to our sensoric, neuronal and cognitive capabilities, all science is equally inevitably reductionist. However, this reductionism differs from traditional reductionism: Firstly, reduction is made consciously and seen as inevitable, secondly, it states that any knowledge in itself is already a reduction. And this reduction is openly admitted: We do not see that we do not see, because we see what we see [1].

In the framework of second order system theories, systems appear as reasonable reduction of complexity [33]: The delimitation of a system from its environment serves as reduction of complexity, or even takes place because of it. Without emerging order (that is, the system) everything can be related to everything else. As a consequence, these relations cannot be utilized anymore, or rather: they are no longer feasible. To summarize, there exists an abstract potential for relations from which—according to narrow conditions—the structure of the system determines those relations that are permitted. The complete number of interdependencies is thereby reduced. This reduction then allows the creation of boundaries between system and environment, and to qualify elements as belonging to the system. The kind of reductions that are chosen results from the relation between system and environment, that is: there exist only limited ways of existence within an environment, which is far more complex than the system itself, thus also limiting the possible choice of system-environment-relations [34], also cf. [35].

The implications and problems that arise from traditional reductionism—being completely different from this special kind of reductionism—are the subject of the following chapter.

9.2 Reductionism

He then has the component parts in hand but lacks, alas! the spirit's band.

Goethe

In today's sciences, reductionism is widespread, and with it thus the assumption that the proceeding splitting of study objects into increasingly smaller parts is a key to the understanding of the whole. This, in turn, is principally based on the assumption of determinism. Determinism is an essential component of Newtonian mechanics: the laws of motion absolutely determine together with the present dynamical state, that is, the position and movement of all particles in the universe, the past and the future of the universe. If all individual components and their behaviour were known, this thus would allow conclusions about the world as a whole. Within determinism, future, present and past are thus logically equivalent as any single one of them determines the other two [36].

Similarly, reductionism is widespread within geomorphology. One example has already been presented in Chap. 6: The study of a single variable in order to determine the equilibrium of the whole system. Reductionism assumes "that even the most complex of systems, when viewed at the 'component level', somehow becomes simpler" [37, p. 137]. However, the question remains unanswered, how the single (or rather: the separated) components can be re-united with other components of the same or even another scale.

The physician Pehr Sällström [38] is aware of this problem. He discusses a painting of Oscar Reutersvärd (cf. Fig. 14) which illustrates a three-dimensional object. At a closer look, the object becomes 'impossible'—it only appears to be three-dimensional. This, however, is only revealed if the object is considered as a whole. If a small section is focused, the peculiarity and oddness of the dimensionality disappears: Any small section seems to be completely coherent. Sällström [38] concludes that the re-composing of single coherent parts can never guarantee the coherence of the whole. This is precisely the core problem of reductionism, which has already been expressed by Aristotle's statement "the whole is more than the sum of its parts". That the putting together of parts does not inevitably (have to) lead to a coherent whole is quite obvious for paintings, but all the more difficult for real phenomena.

Empirical findings from high-energy physics (physics of elementary particles) support these logical reservations towards reductionism (cf. [39]): They show that the so-called elementary particles (e.g. neutrons, protons, electrons)[4] are deprived of elementarity, and even of materiality—they, themselves, are something compound, paradoxically something compound of themselves: They constantly transfer themselves into each other when observed [40, p. 9]. As, depending on the approach (that is, depending on the technique of measurement)[5] the elementary particles appear as wave or as particle [39]. Here, the limitation of reducing the system to increasingly smaller parts can no longer be ignored: It is of utmost importance for the functioning of reductionism that while interacting with other parts one part is not substantially changing its characteristics so much that it can no

[4] Apart from those mentioned above there are numerous other particles with similar characteristics. All of them are summarized as "elementary particles", even though they can hardly be compared with each other. For example, one class of 'elementary particles', the so-called hadrons (i.e. strongly interacting particles) are themselves made up of quarks, so that they are by no means 'elementary' (cf. [39], p. 18f.).

[5] Please also refer to the thoughts on the questionable objectivity of instruments (Chap. 2).

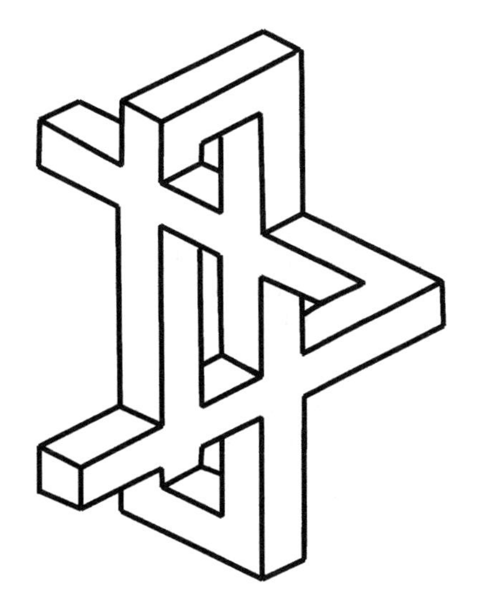

Fig. 9.4 Perspective japonaise no 264 from Oscar Reutersvärd. The focus on one section of the whole veils the realisation that the composite whole does not 'make sense (from [38], modified)

longer be recognized as part. Hence, diversity can no longer be represented by the simple [39] (Fig. 9.4).

Thus, reductionism has a principle weakness. How is it therefore possible that it has such an impressive success story within the sciences? This is at least partly due to one essential characteristic: As it were, reductionism is inevitably crowned with success. Any study object, any system is simply decomposed into ever smaller parts until finally one part becomes explainable and comprehensible (pragmatic reductionism, cf. Riedl [11], p. 85). Additionally, Dürr leads this success story back to the importance of mechanics in our everyday lives: Due to mechanics, the attempt 'to understand something' nearly became equivalent to dissecting something to its parts, i.e. to analysis. This does not pose a problem for all those study objects within which any element can be regarded independently from all the other elements. For example, Bertalanffy [41, 42, p. 147] illustrates this case with a pile of stones or, equivalently, mechanical approaches with which acting forces can be analysed by means of a parallelogram of forces. However, living systems as well as physical systems are profoundly characterized by the interactions of their elements. As these interactions (the reciprocal effects) of the single parts are to be understood [39] this is where the problem arises since these interactions are no longer manifest, or they are changed due to the isolation of the parts. For this dissertation, the theory of dissipative structures probably gives the most concise example. After all, what has happened? When the mechanisms and structures of physical and biological phenomena were reductionistically and linear-kinetically studied after experimental isolation under conditions close to equilibrium, essential characteristics were lost. These were exactly those characteristics of systems far from thermodynamic equilibrium under non-linear conditions, just as in nature, when being coupled to a multitude of dynamic neighbouring processes [43, p. 62].

Hence, this expectation that reductionistically identified causes and effects would be a concise explanation of 'what really is'. It can be termed ontological reductionism and goes further than the pragmatic reductionism [11, p. 85].

Ashby [44] expresses the dilemma of reductionism as follows:

> Science stands today on something of a divide. For two centuries it has been exploring systems that are either intrinsically simple or that are capable of being analysed into simple components. The fact that such a dogma of "vary the factors one at a time" could be accepted for a century, shows that scientists were largely concerned in investigating such systems as *allowed* this method; for this method is often fundamentally impossible in the complex systems [...] there are complex systems that just do not allow the varying of one factor at a time—they are so dynamic and interconnected that the alteration of one factor immediately acts as a cause to evoke alterations in others, perhaps in a great many others. Until recently, science tended to evade the study of such systems, focusing its attention on those that were simple and, especially, reducible [44, p. 5].

One geomorphological example for these fallacies that may arise from the reductionist approach has already been discussed: the description of system equilibria by Bull [45]. Bull tries to explain landscape changes with system equilibria and geomorphic thresholds. In his example, he focuses on a variable in a specific time unit: streambed altitude. The restriction to one variable—without preceding investigation whether it is characteristic for the system state at all—and to one temporal scale—that, equally, might not be characteristic for the system behaviour as a whole—results in the detection of an equilibrium. Hence, from this possibly given characteristic of one element the characteristic of the whole system is inferred. This seems to be logically and physically incorrect. Furthermore, it is yet another example for the mixing of the different levels of physical investigation of systems (also cf. Chap. 5). Altogether, the decades of equilibrium thinking seem to be an expression of a strong reductionist conception of the world: To put it simply, reductionism can be seen as the attempt or the desire to reduce the complex to the very simple (cf. [39]. At the same time—especially in system theoretical approaches—there is no attempt to reach structural simplicity, but rather particularly simple behaviour in time. And the simplest behaviour in time is *temporal invariability* or *temporal conservation* [39].

In the above example, but especially also in the concepts of graded time and steady time, as presented by Schumm & Lichty [46], the extension of reductionist thinking from the material level (within which systems are disaggregated to their 'individual parts') to the temporal dimension can be observed: Obviously, it is assumed that not only space, but also time can be dissected into ever smaller parts that can then be described and understood more easily. This has the advantage (and maybe also the purpose) that linearities and thus causalities arise. Within the next chapter, we will have a closer look at this issue. Subsequently, we will discuss the extent to which this assumption of 'simplicity of the microscopic level' agrees with the 'simplicity of the macroscopic level' of dissipative structures [47].

9.3 Causality

A cause is effect of its own effect, and its effect is its cause.

Ibn' Arabi

Science does not only aim at observing the world and at constructing knowledge about it. Science also and particularly wants to find out causes and principles that presuppose the observed phenomena [2]. It even seems to be an everyday phenomenon that we seek explanations for those things and experiences that catch our attention. Explanations in the sense of 'how such thing causes such': The question of the relationship of cause and effect (also cf. Egner [3]), the question of 'why' [48]. Whilst doing so, we expect that we can understand all phenomena by reducing them to their components (cf. [11, p. 85]). However, this is problematic as cause and effect can easily be mixed up, or the cause and effect-relation is wrongly constructed. Furthermore, circular causalities—those cases in which the effect acts back on the initial cause—are often simplified to linear causalities. For example, the circular causality that the orbit of the moon determines the tides, and the tides in turn affect the orbit of the moon, becomes the linear cause and effect-relation that the tides are due to the moon (cf. [11, p. 69]). Quite simply, causality is a challenge [48] and even darker by far and more difficult than philosophers generally like to admit [13, p. 133]. This is partly due to the fact that the notion of cause is highly unspecific:

> It commits us to nothing about the kind of causality involved nor about how the causes work. Recognising this should make us more cautious about investing in the quest for universal methods for causal inference [49].

After all, following Russell's fundamental criticism of causality it seemed to be settled for the exact sciences (cf. [13]). According to Russell, causal dependency is a dangerous concept, as in the end, cause and effect often cannot be distinguished, and both statements are equally true and wrong [13]. Causality is based on the idea that for any event there must be a cause without which the event would not exist. Prigogine and Stengers [50, p. 4f] even go a step further: Causality means the equivalence of complete cause and full effect. This does not express anything but that anything can be explained and that nothing happens without a cause. The principle of sufficient reason does not only postulate the existence of a cause, but also the equivalence of cause and effect. It was Leibniz, who in his theory of movement, i.e. dynamics, discovered the richness of this principle. Indeed, the work of Galilei and Huyghens did not primarily make sense to the chain of cause and effect, but to the equality of cause and effect. On an inclined surface, in the ideal case a body gains exactly that amount of velocity that brings it back to its initial height. What has been 'consumed' by the falling equals that which has been 'produced' by the falling. This equality is confirmed by the possibility to reverse the process: As the moving body 'consumes' the generated velocity it can retrieve what it has lost, that is, its initial height. These thoughts were followed by the conception of (kinetic and potential) energy and allow for putting the equal sign between quantities: in between mgh, the 'lost' quantity due to the falling from h to zero, and $m \times v^2/2$, the quantity gained by the fall of a body that resulted in the velocity v.

Russell refrained from causality for empirical reasons: According to him, it is simply impossible to correctly and completely reproduce the initial conditions adequately. Even if this had been achieved, it would be practically impossible that the conditions recurred in exactly the same way [13]. Thus, the cause vanishes within a network of pre-conditions [48]. In this sense the farewell to causality also means a farewell to the ideal of omniscience, a conclusion which was shared by Prigogine and Stengers roughly 40 years after Russell's pamphlet.

For Russell, another reason to excoriate causality was that any cause and effect-relation claims a certain uniqueness (cf. [13], p. 135f). However, according to Russell there is no reason to assume that there is no other possible interaction or interplay of pre-conditions that could lead to exactly the same effect. The contrary is also possible: The same causes may lead to different effects. Geomorphologists are aware of these complications, but it is criticised as the lack of an explicit understanding of cause and effect:

> We know, for example, that Holocene climate changes were severe enough to upset the delicate balance at the surface. What is confusing is that diverse geomorphic responses often result from the same climatic trends. Thus, even for the very recent past we commonly are unsure of cause and effect (Ritter et al. 1995, p. 5).

Until now, a geographical epistemological and methodological discussion of causality has rarely taken place,[6] though such a discussion is literally intruding in view of the determinism-centred geomorphological system theory. After all, one of the main questions is how such (causal and linear) determination of system behaviour can be assumed within a system theory, which, per definition, assumes reciprocation and positive and negative feedbacks that are contradicting linear relations. For example, even the most classical geomorphological question of process and form results in the difficulty of what is cause and what is effect. In endless recursive loops, would form not be its own cause? Schumm and Lichty [46] posed similar questions and concluded that a cause can become an effect and vice versa as soon as systems are considered dynamically and not statically. Thus, system behaviour can no longer be described and understood with linear approaches. Reciprocal dependence of form and process constitute a circularity that is theoretically pictured with the theory of dissipative structures of Prigogine.

It can be concluded that the approach to causality is yet another, important difference between mechanistic and thermodynamic concepts: A dynamic system is studied by focusing interacting objects. If *a* acts with a defined force and in a defined way on *b*, (with known framework conditions) the behaviour of *b* can be fully determined and traced back to the action of *a*. However, even within classical system theory there exists a causal-theoretical problem: the postulated reciprocal

[6] An essential exception is the already cited work of Joachim Rathmann [48].

effects violate the prerequisite of asymmetry.[7] Additionally, the notions of 'reciprocation' (or 'interaction') are more or less openly contradictory [48].

Things are further complicated if, instead of linear systems, non-linear, self-organised systems are considered. For example, it can be questioned how self-steering and causality agree with each other. Self-steering can be regarded as some kind of *causa finalis*, thus explaining the present by means of the future.[8] This in turn implies a purposefulness or teleology that is generally only ascribed to conscious systems and therefore brings with it several philosophical problems (cf. [48]. Finally, in chaotic systems, causality is replaced by probabilities. If then emergent phenomena are included, these difficulties are further enhanced as emergence excludes causal reduction to single elements.

Consequently, for the study of thermodynamic systems a completely different approach has to be utilised, as within them at best correlations[9] can be observed. These, in turn, may change with the intensity of the processes (cf. e.g. Prigogine and Stengers [50] for details on this aspect). Here, statements on cause and effect are not imperative: For any elementary, structure-building process, previously unimportant external (environmental fluctuations) or internal (statistic noise on microscopic level) disturbances may suddenly gain crucial importance by governing timing and direction of the processes. For self-organisation, coincidence becomes essential [51].

Moreover, if autopoietic systems are assessed, these problems are aggravated: self-reference has to be taken into account. Self-reference describes the circularity of cause and effect. Thus, it cannot be understood and captured by the theories on causality as the effect may precede the cause [48]. This case is absolutely excluded by the definition of causality that states that the effect has to follow the cause.

To conclude, first order system theories, but especially second order system theories pose a challenge for causal theories. How these challenges can be met, or whether we are approaching a second farewell to causality after Russell's pamphlet 'On the notion of cause' in 1912/1913 currently remains unclear. Rathmann [48] suggests a possible solution by means of the abolition of the dichotomy of *causa efficiens* and *causa finalis*, resulting in an abolition of the separation of natural and social sciences. The latter has already been demanded by Prigogine, albeit for other reasons. Unlike Rathmann, Prigogine and Slenges [50] consider the farewell to causality and thus the farewell to the ideal of omniscience as a chance rather than a problem: If the ideal of omniscience, borne by the notion of the

---

[7] The relation of cause and effect is asymmetric as the cause always has to precede the effect (see [48]).

[8] Aristotle distinguished four different causes: Causa materialis (material cause), causa formalis (formal cause), causa efficiens (efficient cause), and causa finalis (final cause). For more detailed explanations of the Aristotelian understanding of causality see e.g. Riedl [11] and Egner [3].

[9] Correlations only allow insights on (linear) relations between different variables, but they do not express whether these variables are causally linked. Probably, one of the most popular examples is that of decreasing birth rates along with a reduction of the stork population. This correlates, but does not imply any causal relation.

dynamic trajectory, is incompatible with the instability of dynamic systems; if the principle of sufficient cause ceases to be, as Leibniz has put it, the Ariadne's thread in the labyrinth of phenomena, this does not imply a defeat, or a renouncement. Quite the contrary, it is a beginning, a point where new questions arise [50].

In this sense, geomorphology is right at the point of a new beginning. We have reached a 'bifurcation point' at which we can determine in a self-organised and self-referential manner which future direction we choose. Entirely in accordance with the observation theory as presented in Chap. 2, this choice of direction is not a choice of right or wrong. What it does mean, however, is that geomorphology now has the chance—or the opportunity—to deliberately start to discuss the foundations of geomorphological research. This, indeed, can be a chance, a space of opportunity, rather than a threat.

References

1. Fuchs P (1992) Niklas Luhmann-beobachtet. Eine Einführung in die Systemtheorie. Westdeutscher Verlag, Opladen, p 219
2. Peschl MF (2001) Constructivism, cognition, and science–an investigation of its links and possible shortcomings. Found Sci 6(1–3):125–161
3. Egner H (2010) Theoretische Geographie. Wissenschaftliche Buchgesellschaft, Darmstadt, pp 144
4. Russell B (1952) Mystik und Logik. In: Russell B (ed) Mystik und Logik. Philosophische Essays. Humboldt, Wien, Stuttgart, pp 5–35
5. Maturana HR, Varela FJ (1984) Der Baum der Erkenntnis. Die biologischen Wurzeln des menschlichen Erkennens. p. 280
6. Varela FJ (1997) Erkenntnis und Leben. In: Simon FB (ed) Lebende Systeme: Wirklichkeitskonstruktionen in der systemischen Therapie. Suhrkamp, Frankfurt/Main, pp 52–68
7. Riegler A (2001) Towards a radical constructivist understanding of science. Found Sci 6(1–3):1–30
8. Seckel Al (2006) The ultimate book of optical illusions. Sterling, New York, p 367
9. Wardenga U (2001) Zur Konstruktion von Raum und Politik in der Geographie des 20.Jahrhunderts. In: Reuber P, Wolkersdorfer G (eds) Politische Geographie: Handlungsorientierte Ansätze und Critical Geopolitics. Heidelberger Geographische Arbeiten, Heidelberg, pp 17–31
10. Schmidt SJ (1987) Der Radikale Konstruktivismus: Ein neues Paradigma im interdisziplinären Diskurs. In: Schmidt SJ (ed) Der Diskurs des Radikalen Konstruktivismus. suhrkamp, Frankfurt/Main, pp 11–88
11. Riedl R (1981) Die Folgen des Ursachendenkens. In: Watzlawick Paul (ed) Die erfundene Wirklichkeit. Wie wissen wir, was wir zu wissen glauben? Pieper, München, Zürich, pp 67–90
12. Russell B (1952) Über die wissenschaftliche Methode in der Philosophie. In: Russell Bertrand (ed) Mystik und Logik. Philosophische Essays. Humboldt, Wien, Stuttgart, pp 98–124
13. Russell B (1952) Über den Begriff der Ursache. In: Russell B (ed) Mystik und Logik. Philosophische Essays. Humboldt, Wien, Stuttgart, pp 181–208
14. von Foerster H (2006) Sicht und Einsicht. Versuche zu einer operativen Erkenntnistheorie. Carl-Auer, Heidelberg, p 233

15. Maturana HR, Varela FJ, Frenk SG (1982) Größenkonstanz und das Problem der Wahrnehmungsräume. In: Maturana HR (ed) Erkennen: Die Organisation und Verkörperung von Wirklichkeit. Ausgewählte Arbeiten zur biologischen Epistemologie. Wissenschaftstheorie, Wissenschaft und Philosophie. Vieweg, Braunschweig/Wiesbaden, pp 81–87
16. Schmidt SJ (1982) Einladung, Maturana zu lesen. In: Maturana HR (ed) Erkennen: Die Organisation und Verkörperung von Wirklichkeit. Ausgewählte Arbeiten zur biologischen Epistemologie. Wissenschaftstheorie, Wissenschaft und Philosophie. Vieweg, Braunschweig/Wiesbaden, pp 1–10
17. Maturana HR (1982) Kognitive Strategien. In: Maturana HR (ed) Erkennen: Die Organisation und Verkörperung von Wirklichkeit. Ausgewählte Arbeiten zur biologischen Epistemologie. Wissenschaftstheorie, Wissenschaft und Philosophie. Vieweg, Braunschweig/Wiesbaden, pp 297–318
18. von Glasersfeld E (1992) Why I consider myself a cybernetician. Cybern Human Knowing 1(1):21–25
19. Maturana HR (1982) Biologie der Kognition. In: Maturana HR (ed) Erkennen: Die Organisation und Verkörperung von Wirklichkeit. Ausgewählte Arbeiten zur biologischen Epistemologie. Wissenschaftstheorie, Wissenschaft und Philosophie. Vieweg, Braunschweig/Wiesbaden, pp 32–80
20. Maturana HR (1982) Biologie der Sprache: die Epistemologie der Realität. In: Maturana HR (ed) Erkennen: Die Organisation und Verkörperung von Wirklichkeit Ausgewählte Arbeiten zur biologischen Epistemologie. Wissenschaftstheorie, Wissenschaft und Philosophie. Vieweg, Braunschweig/Wiesbaden, pp 236–271
21. Maturana HR (1982) Erkennen: Die Organisation und Verkörperung von Wirklichkeit. Ausgewählte Arbeiten zur biologischen Epistemologie. Wissenschaftstheorie, Wissenschaft und Philosophie. Vieweg, Braunschweig/Wiesbaden, p 322
22. Maturana HR (1994) Was ist Erkennen? Piper, München, p 244
23. Spencer-Brown G (1972) Laws of form. Julian Press, New York, p 216
24. Spencer-Brown G (1996) Wahrscheinlichkeit und Wissenschaft. Carl Auer, Heidelberg, p 142
25. von Glasersfeld E (1996) Radikaler Konstruktivismus. Ideen, Ergebnisse, Probleme. Suhrkamp, Frankfurt/Main, p 375
26. von Foerster H (1960) On self-organizing systems and their environments. In: Yovits MC, Cameron S (eds) Self-organizing systems. Pergamon Press, New York, pp 31–50
27. von Foerster H (1984) Observing systems. Intersystems Publications, Seaside, p 331
28. von Foerster H (2002) In jedem Augenblick kann ich entscheiden, wer ich bin. In: Pörksen B (ed) Die Gewissheit der Ungewissheit. Gespräche zum Konstruktivismus. Carl Auer, Heidelberg, pp 19–45
29. von Glasersfeld E (2002) Ernst von Glasersfeld im Interview mit Reinhardt Voss. "… es ist eine anstrengende und vor allen Dingen ungemütliche Sache". In: Voss R (ed) Unterricht aus konstruktivistischer Sicht. Die Welten in den Köpfen der Kinder. Luchterhand, Neuwied, pp 26–32
30. Luhmann N (1997) Selbstreferentielle Systeme. In: Simon FB (ed) Lebende Systeme. Wirklichkeitskonstruktionen in der systemischen Therapie. Suhrkamp, Frankfurt/Main
31. von Glasersfeld E (2002) Was im Kopf eines anderen vorgeht, können wir nie wissen. In: Pörksen B (ed) Die Gewissheit der Ungewissheit. Gespräche zum Konstruktivismus. Carl Auer, Heidelberg, pp 46–69
32. Glanville R (2001) An observing science. Found Sci 6(1–3):45–75
33. Luhmann N (1992) Die Wissenschaft der Gesellschaft. Suhrkamp, Frankfurt a.M., p 732
34. Luhmann N (1976) Komplexität. In: Luhmann N (ed) Handwörterbuch der Organisation. Stuttgart, pp 939–941
35. Luhmann N (1975) Komplexität. Soziologische Aufklärung. Aufsätze zur Theorie der Gesellschaft (Band 2). Opladen, Westdeutscher Verlag: 204–220
36. Penrose O (1984) Improving on Newton. Nature 310:341

37. Favis-Mortlock D, De Boer D (2003) Simple at heart? Landscape as a self-organizing complex system. In: Trudgill S, Roy A (eds) Contemporary meanings in physical geography. Arnold, London, pp 127–172
38. Sällström P (1992) The possibility of the impossible. Cybern Human Knowing 1(1)
39. Dürr H-P (1986) Neuere Entwicklungen in der Hochenergiephysik-das Ende des Reduktionismus. In: Dress A, Hendrichs H, Küppers G (eds) Selbstorganisation. Die Entstehung von Ordnung in Natur und Gesellschaft. Piper, München, Zürich, pp 15–34
40. Dress A, Hendrichs H, Küppers G (eds) (1986) Selbstorganisation. Die Entstehung von Ordnung in Natur und Gesellschaft. Piper, München, Zürich, p 234
41. von Bertalanffy L (1950) The theory of open systems in physics and biology. Science 111(2872):23–29
42. von Bertalanffy L (1950) An outline of general systems theory. Br J Philos Sci 1(2):134–165
43. Hess B, Markus, M (1986) Chemische Uhren. In: Dress A, Hendrichs H, Küppers G (eds) Selbstorganisation. Die Entstehung von Ordnung in Natur und Gesellschaft. Piper, München, Zürich, pp 61–80
44. Ashby WR (1956) An introduction to cybernetics. Chapman & Hall, London, p 295
45. Bull WB (1991) Geomorphic responses to climatic change. Oxford University Press, New York, Oxford, p 326
46. Schumm SA, Lichty RW (1965) Time, space, and causality in geomorphology. Am J Sci 263(February):110–119
47. Jantsch E (1979) Die Selbstorganisation des Universums. Vom Urknall zum menschlichen Geist. Hanser Verlag, Darmstadt, p 464
48. Rathmann J (2008) Kausalität in der Systemtheorie: Ein Problemaufriss. In: Egner H, Ratter BMW, Dikau R (eds) Umwelt als System - System als Umwelt? Systemtheorien auf dem Prüfstand. oekom, München, pp 55–71
49. Cartwright N (1999) The dappled world. A study of the boundaries of science. Cambridge University Press, Cambridge, p 247
50. Prigogine I, Stengers I (1990) Entwicklung und Irreversibilität. In: Niedersen U, Pohlmann L (eds) Selbstorganisation und Determination. Selbstorganisation. Jahrbuch für Komplexität in den Natur-, Sozial- und Geisteswissenschaften. Duncker & Humblot, Berlin, pp 3–18
51. Pohlmann L, Niedersen U (1990) Dynamisches Verzweigungsverhalten bei Wachstums- und Evolutionsprozessen. In: Niedersen U, Pohlmann L (eds) Selbstorganisation und Determination. Selbstorganisation. Jahrbuch für Komplexität in den Natur-, Sozial- und Geisteswissenschaften. Duncker & Humblot, Berlin, pp 63–82

Chapter 10
Practical Implications

> *I have no doubt that, in reality, the future will be much more surprising than anything that I can imagine.*
>
> J.B.S. Haldane

10.1 Prediction and System Control

Especially within today's research landscape many disciplines encounter the demand that their research needs to be of relevance for society. After all, the amount of research funds allocated is determined by this aspect. Science, or rather the knowledge generated by science, is seen as an instrument for better predictions and for more effective environmental management and exploitation (also cf. [1]). For this, a genuine scientific interest in a deeper understanding of phenomena is no longer necessary. Instead, the focus shifts to functional relations and to regularities of dynamics: Causality trumps (see above), as well as efficiency. On the one hand, causalities allow for prediction (if this happens, then that follows), and on the other hand, causalities allow for management (if this is supposed to happen, I have to arrange this). Thus, the management of geomorphic systems—be they river systems, slope systems, or coastal systems—is based on the initial assumption that system control is principally possible. According to Krohn and Küppers [2], the basic idea of system control is to influence a system such that a certain target idea is reached. Thus, a model of reality is designed that illustrates the actual state as well as the 'normal' (that is, the assumed natural) system behaviour, and an action plan is supposed to lead to the given target. This way of thinking implies several preconditions: For example, the observer is (or has to be) capable of fully accessing and understanding the behaviour of the system, including possible feedback loops. Furthermore, the system can be fully described by the framework conditions and the elements of the system, and it has to react linearly to external influences. Additionally, an effect has to be relatable to a cause, and thus a certain impact (including attempts to control the system) is supposed to lead to a certain reaction. Utilizing the geomorphological ductus: An input is necessarily followed by a specific output. Last but not least, system management and control needs to assume that the present observation of systems is sufficient for understanding

K. von Elverfeldt, *System Theory in Geomorphology*,
Springer Theses, DOI: 10.1007/978-94-007-2822-6_10,

systematic relations, and that the presently observed relations do not change with time (uniformitarianism). Thus, from the current situation, the past as well as future incidents are inferred. It principally holds that only self-regulating systems can be controlled, as this self-regulation can be manipulated. However, these (geomorphic) systems differ significantly from e.g. machines, that is, man-made systems, as for the latter the internal procedures are known, plus machines are designed to fulfil a certain purpose. In contrast, the chances to control a geomorphic system are few, and they are fewer the lesser initial and framework conditions can be determined, and the more specific the system (cf. [3, 2]).

But what does this mean for the handling of systems far from (thermodynamic) equilibrium? These show self-organisation and non-linearity, and causalities are not necessarily found. After all, for these systems the above premises no longer hold—one minor irritation, one attempt to control can not only principally lead to more than one effect, but can also have different effects at different times. Basically, this strongly questions the assumption that geomorphic systems can be planned and controlled. And don't we know this phenomenon from practice? After all, it becomes especially (and tragically) evident that geomorphic systems cannot be controlled whenever natural disasters occur—again and again it can be observed that risk management fails (as was the case for the Hurricane Katrina, also cf. [4]). Less tragically and drastically, the same phenomenon of lack of control and correct prediction shows ex-post for the construction of hydropower plants (with the unintended consequences of river bed incision and the resulting problem chain), for river bed regulation (and the resulting problems of coastal erosion, as e.g. at the river Weser)[1], for the failure of 'secure' hillslopes, etc. Another example is provided by the experiment on dike stability at the University Karlsruhe (already mentioned in Sect. 4.2): Though dike failure was predicted, the failure did not occur. Insights from the self-organized criticality of slopes (cf. e.g. [5]) hint at the same direction. They suggest that pore water pressure cannot be the only criterion for the explanation of slope failure. For example, this could add new insights to the catastrophe of Vaiont in 1963, where a hillslope started to move when the lake was dammed. Slope failure finally caused a flood wave, leading to 2,500 casualties (cf. [6, 7]). When the movement of the slope was noticed, an attempt was made to control the movement by adjusting the lake level. However, from a certain point onwards, the slope did not react as expected and collapsed into the lake. The theory of dissipative structures offers an explanation for this behaviour: before failure, the slope reached the instability threshold, thus from then acting in a self-organised manner and therefore widely independent from external factors and influences. However, in most cases, such unforeseen consequences are ascribed to a lack of system knowledge instead of acknowledging the possibility that these unforeseen consequences are immanent to self-

[1] This is only true forthe first constructions of water plants, when the above mentioned consequences had not been foreseen. In the meantime, the interrelations are better known, and sometimes the consequences are accepted tacitly, since the profit exceeds the costs. The same is true for the coastal erosion caused by channel incision.

organised systems and thus cannot be avoided by further accumulation of knowledge. After all, when system state follows upon system state, that is, any state is a consequence of the preceding one, prediction will rarely prove successful. It will rarely be successful, even if the mechanism of the system evolution is known, if the systems are deterministic and disturbances are absent—even these systems can show chaotic behaviour. As in this case their history, i.e. their sequence of states depends fully on the initial conditions (which are unknown), their behaviour is unpredictable [2, p. 114f].

The problem seems to be well-known, yet it is completely different from the previous controlling and planning problems: Even linear systems (or systems thought to be linear) are highly complicated due to their immensely large number of variables. Mostly, the reflex was to strive for a 'more' of data that can be used as parameters for models—a 'black box' should become a 'grey box'. In the case of self-organised and/or self-referential systems, however, it becomes obvious that this is no solution: A small number of variables is quite sufficient to cause complexity or chaos, and even 'more realistic' modelling does not result in a better prediction. Basically, non-linear, self-organised systems thus sound the death knell for induction. After all, even the comparison and repeated observation of similar systems is not helpful, since the slightest deviations of the initial conditions are capable of initiating completely different system behaviour.

Furthermore, it is increasingly acknowledged that, in a manner of speaking, the present is too short to explain the past or the future. Present system states or, equally, those states that prevailed before mankind, cannot be regarded as 'the normal', or the ideal state, to which the system can or should be brought back or within which it should be kept. Additionally, in complex systems, small deviations of the initial conditions can eventually result in completely different system behaviour. Even if we had relatively good knowledge of one specific system, this knowledge could not necessarily be transferred to a similar system.

As was detailed in Sect. 4.2, close to bifurcation points, dissipative structures are characterized by instability, whilst they are stable and regular when outside these areas of instability (cf. [8]). When a system reaches a bifurcation point, the choice of the future path is dictated by chance. At the same time, the describing equations can be completely deterministic (chaotic attractor). Any point of this attractor represents a possible preliminary final state of the system. However, this preliminary final state is not necessarily stable—smallest changes can bring the system to a completely different state. Thereby, the system cannot distinguish whether the disturbances are purposeful or coincidental. Any noise, any fluctuation, and any intervention possess a momentum of indeterminacy [9, p. 80]. Whilst a single event is absolutely meaningless in a stable state, it can—whilst in an instable state—bring the system to a completely different state. This determining effect of *single elements* completely contradicts the idea that (safe) predictions can be made, especially as the theory of probability fails here: For the correct probabilistic prediction of a specific event is close to zero [10, p. 36].

Additionally, during controlling, that is, whilst the respective operations are actualised, a billion other things are already taking place, which—as they are simultaneous—can neither be known nor influenced ([11, p. 12, cited according to [4]). Against this background a 'direct control' of system behaviour seems impossible and illusionary. Furthermore, any bifurcation point means at least a halving of the prediction probability[2]. Based on this, a macroscopic indeterminacy relation can be calculated (cf. [9, p. 80]), which with n bifurcations corresponds to $p \times 2n < 1$. Here, probability decreases exponentially, from which a principal horizon of prediction results. Any prediction is further complicated by the fact that humans are constantly acting on the natural systems, and they are acting back on us. Simply speaking, we are not standing outside the world, where we could take a *locus observandi* that places us externally, so that we could observe literally everything.

This acknowledgement of the principally limited predictability strongly contrasts the traditional realistic notion of science: For example, Heinrich Hertz [12] introduced his work "Die Prinzipien der Mechanik in neuem Zusammenhange dargestellt" with the words:

> "Es ist die nächste und in gewissem Sinne wichtigste Aufgabe unserer bewussten Naturerkenntnis, dass sie uns befähige, zukünftige Erfahrungen vorauszusehen um nach dieser Voraussicht unser gegenwärtiges Handeln einrichten zu können" ([12, p. 1f], cited according to [10, p. 25]) ("It is the imminent and, in some sense, most important task of our conscious knowledge of nature that it should enable us to predict future experiences, so that we can adjust our present acting accordingly").

Considering the theory of non-equilibrium thermodynamics, this "most important task"—and for many geomorphologists prediction is still the most important task—has to be rethought. For the question is now, how can the principally low predictability of future system behaviour be dealt with, especially as changes within non-linear systems may well be catastrophic (cf. [10])? One consequence is that we cannot protect ourselves against surprises and coincidences [13]. How can these systems be controlled? A planning dilemma results: Within modern societies, plans for the future encompass almost all spheres of life and nearly all world religions. They span periods stretching across the range of decades. The plans have increased with the same amount as the technical and social innovation rhythms became faster and, more importantly, unforeseeable—nearly everywhere, planning is chasing after the actual developments. This rather strongly indicates that planning rhetoric is based on deficient planning competence. It can thus be questioned in how far this development is a consequence of either insufficient planning instruments or of excessive planning aspirations. Strong doubts remain whether an improvement of planning instruments should be aimed at, as the insights from the theory of self-organised systems rather point to a principal limit of the feasibility of better predictions of complex system behaviour, especially if the reactions to control attempts are to be included. So far, from the analysis of self-referential, paradoxical systems it

---

[2] At those bifurcation points where the system can choose between more than two pathways, the probability is accordingly reduced by a greater amount.

remains unclear how to conduct planning that includes itself, or its own consequences within the planning [2, p. 109].

This planning dilemma appears even more drastically, if the concept of autopoiesis is applied to approaches of planning and control: The concept of autopoiesis does not offer any control concepts (also cf. [14]). Without such a concept, however, there is no alternative offer of a changed practice, possibly without the 'collateral damages', which we are already used to. With the concept of autopoiesis, the problem of (not) controlling systems cannot be solved directly—but instead, it offers the possibility to generate a new and broader understanding of system control. If—instead of the allopoietic perspective with its focus on input–output-relations and causalities—an autopoietic perspective is chosen, the organisation and the structure of the system is focussed instead. According to Bühl [14], system control theories benefit from an autopoietic perspective by gaining a better control structure. However, he does not offer a specific concept.

System control and prediction of autopoietic systems is further limited by the fact that they, per definition, are not determined by their (relations to their) environment. This aspect is central for the application of system theory. The established assumption that systems can be specifically influenced from the outside is crucial for any prediction of future system behaviour (cf. [15]). At the same time, even deterministic systems are not necessarily predictable, as this would mean that their changes of state can be calculated. For any prediction it is assumed that the system is a coherent, deterministic system, and that its constellations can thus be completely calculated, that is, predicted. Whenever this is not possible, any prediction has to fail. Thus, determinism neither equals predictability, nor does the lack of predictability hint to non-deterministic systems—it only proves that certain calculations simply cannot be made [16, p. 224].

According to these scientific theories (long-term) prediction is thus largely impossible (cf. [2]). Does this ultimately imply that we have to give up any thought of system control and management? This does not seem to be an attractive alternative, especially as non-controlled systems may also become instable and show catastrophic behaviour (also cf. [13]). Apparently, it is a major problem that the planning and control of systems is often based on the assumption of rigid process structures, and this structure is (linearly) projected into the future. Thus, the alternative seems to be flexibility rather than laissez-faire or 'just doing nothing'. After all, giving up on any attempts of system control and management would, as a consequence, question all societies, which seems to be rather absurd. Instead, I tend to understand the above thoughts as warning note and invitation at the same time: As warning that successful system control is already improbable due to the diversity and high number of involved social part-systems—ranging from the science system to the political, the legal, and the economic system. Therewith result several equitable and equally 'true' perspectives, thus providing a surplus of possible solutions [17, p. 151]. But equally it can be understood as invitation to search for pathways of system control that are just as successful as control can be. After all, scientific predictions do not always fail, which paradoxically was illustrated by hurricane Katrina: Already one year earlier, a scenario existed that relatively exactly predicted the damages ([18], also cf. [17]). However, exactly the same scenario revealed the other side of predictions: general fallibility. While the prediction of the consequences of a category 5-hurricane were precise, the scenario wrongly assessed and predicted response measures. These worked for the scenario but failed miserably during the event itself (cf. [19]).

Adaptive management seems to be a promising concept [20]. Here, system behaviour is focussed, that is, the evolution over time as well as the self-regulation or -organisation. This type of planning constantly adapts itself to changed conditions, and only utilizes and generates short-term predictions, thus becoming a management (and not just a planning) approach [13]. The classical steps of target values, models, and action plans are substituted by the definition of target corridors (instead of target states). Scenarios that are based on defined and fixed data correlations and on a specified model simply offer less flexible managing measures.

If the idea to view non-linear thermodynamic systems as self-referential proved viable, in consequence we could no longer claim absolute security. Similar to the administration of medicines, it would be obvious that a certain number of measures has side effects—after all, this estimation already proves true quite often. The estimation of unintended and adverse consequence could then be a main objective of applied geomorphological natural hazard research, just as we already know it from package inserts. Consequently, the failure of planning and control measures would simply be normal, as a universal solution would neither be expected nor strived for.

Basically, the low controllability of autopoietic systems is already transmitted linguistically. For example, Luhmann states that society is divided functionally in part-systems and not in sub-systems. Regardless of whether one thinks that Luhmann's theory is 'realistic' or 'feasible' or not, he is very precise and consistent linguistically. Especially within the context of operative closure, self-reference, and system control the fundamental difference between sub-systems and part-systems and thus the fundamental difference between first and second order system theories becomes obvious: A hierarchic structure expresses a 'top' and a 'bottom', in its most extreme case in a sense of a 'chain of command'. Within first order system theoretical approaches, the hierarchic structure implies both the presence of determinism and controllability: a superordinate system is capable of influencing the functioning and the behaviour of a subordinate system. Along this functional chain steering becomes possible, whilst it is unthinkable for operatively closed, functionally differentiated systems (cf. [17]).

Thus, it depends on the system theoretical assumptions whether or rather, to what extent, the steering and control, planning, and/or management of systems can be suspected (also cf. [17]). If first order system theories are taken as basis, as is the case within geomorphology, system control does not pose a fundamental problem due to environmental determinacy and linear cause-and-effect-relations. Existing problems can easily be referred to as lack of knowledge and/or human failure. However, as soon as the theory of dissipative structures or the concept of autopoiesis is applied, the idea of the controllability of geomorphic systems cannot be maintained any longer. Some of the reasons for this are identical (complexity, non-linearity, self-organisation, structural determinacy, coincidence, and the influence of the history of the system), others do not apply to both approaches, or have not been yet been investigated (e.g. the aspects of self-reference, operative closeness, and autopoiesis). Still, as at least the theory of dissipative structures can most probably be applied to geomorphic systems, this does imply that the management of geomorphic systems has to be rethought. Specific strategies still have to be developed.

However, it is certain that rethinking, thinking differently, thinking of new ways does mean at least: flexible and adaptive. As an implication of a geomorphological second order system theory, the connectivity of the theoretical foundation towards human geography as well as natural sciences can be expected to increase. This seems to be a positive implication, as the scope of solutions could be extended. The potential of this development is shown in the next chapter.

10.2 Inter- und Intradisciplinary Connectivity

Within the chapter on causality I already questioned whether the separation of the sciences into humanities, social sciences, and natural sciences is reasonable. This or similar questions have already been posed by many scientists, among them Prigogine (e.g. [21]): Why should there be more than one science if there is only one world that forms the research object of all sciences? Yet another Nobel laureate, Erwin Schrödinger (1887–1961), considers the pursuance of universal knowledge as the actual objective of science and concludes that some scientists simply have to take the risk of completely disgracing themselves:

> A scientist is supposed to have a complete and thorough knowledge, at first hand, of some subjects and, therefore, is usually expected not to write on any topic of which he is not a master. This is regarded as a matter of *noblesse oblige*. For the present purpose I beg to renounce the *noblesse*, if any, and to be freed of the ensuing obligation. My excuse is as follows: We have inherited from our forefathers the keen longing for unified, all-embracing knowledge. The very name given to the highest institutions of learning reminds us, that from antiquity to and throughout many centuries the *universal* aspect has been the only one to be given full credit. But the spread, both in width and depth, of the multifarious branches of knowledge during the last hundred odd years has confronted us with a queer dilemma. We feel clearly that we are only now beginning to acquire reliable material for welding together the sum total of all that is known into a whole; but, on the other hand, it has become next to impossible for a single mind fully to command more than a small specialized portion of it. I can see no other escape from this dilemma (lest our true aim be lost for ever) than that some of us should venture to embark on a synthesis of facts and theories, albeit with second-hand and incomplete knowledge of some of them—and at the risk of making fools of ourselves. So much for my apology [22, p. 1].

Geography is a discipline, which claims to build a bridge between social and natural sciences. It could be argued that such a bridging discipline is central to the sciences, especially with respect to the 'big questions' of the twenty first century, such as e.g. climate change. Statements such as those by Prigogine and

Schrödinger thus should be grist to the mill of geography. According to this logic, geography should consequently be at the centre of all sciences. However, it remains rather at the sidelines. Furthermore, the bridging can already be questioned discipline-internally, if the far-reaching lack of communication between human and physical geography is considered. This is hard to explain, and—even more important—can this ever be changed?

An explanation of the separation of geography into two part-disciplines would exceed both the scope and the focus of this work. However, some reasons can be inferred from the previous chapters: Although the vocabulary in both part-disciplines might (possibly) be similar in some fields, the same vocabulary might refer to completely different things and issues (cf. [23]). This is, for example, demonstrated by the term 'system'. Another reason for the separation and lack of communication are differing epistemological premises: Whilst physical geography can be largely attributed to realistic positions, human geography largely takes anti-realistic positions.

The question of whether the speechlessness and the *de facto* separated research within geography could be changed has to be preceded by a statement on why one should want this to happen. I argue that the separation of geography is neither reasonable against the background of research purposes (referring to the 'big questions of the twenty first century') nor against the background of science politics. But this does not imply that geography has to re-invent itself. However, a foundation should be found upon which both geographies can collaborate. For this, three paths seem to be possible in principal:

1. On the basis of joint projects. However, it has to be questioned how a mutual understanding of the different basic assumptions and epistemological premises could be reached. Furthermore, which methods should be applied, if the research objects differ fundamentally? This problem also applies to the next point, the
2. development of common methods. Common methods require at least a partly identical or similar methodology, which, however, does not apply to geography.
3. A rapprochement on theoretical grounds. However, for the time being, it remains unclear, which theory could suit this rapprochement.

The first two approaches seem to be impracticable for several reasons. Firstly, the differing epistemological foundations result in different research interests—the questions are simply fundamentally different. Secondly, those fields within which cooperation seems reasonable and within which this is already taking place (e.g. the question of natural hazards and risks) will not have a signal effect on the whole discipline. Thus, a 'boom' in this direction cannot be expected on this basis.

From my point of view, second order system theories offer the only way to generate common research questions and scientific vocabulary for cooperation. The aspect has already been stressed (Sect. 7.1) that the theory of dissipative structures and the concept of autopoiesis have far-reaching commonalities. Furthermore, they offer the potential to serve as background theories for the tackling of shared research problems, as they are applicable within natural sciences (physics, chemistry, biology) as well as social sciences (sociology, psychology, human geography). They have the further advantage that they offer a 'great narrative', having already proved their potential within other disciplines, and being obviously viable. Thus, they can serve as foundation for a broad, theoretically-based geographic approach, built on a shared, sustainable language for a common geography.

Of course, the attempt to get to grips with such 'great narratives' and (even worse?) with 'the other side of geography' is a time-consuming challenge. Furthermore, it poses a certain inestimable risk to thoroughly make a fool of oneself, just as Schrödinger pointed out. However, the advantages prevail, as:

- We may reach a previously unknown scientific connectivity of the two geographies.
- At the same time, it strengthens geography in its relation to its neighbouring disciplines such as physics or sociology. So far, geography is widely stimulated by these disciplines (and others), but has only few concepts to offer in turn.
- None of the two part-disciplines has to take over a theory from the respective 'other setting' and thus—perceived or actually—become subordinate.

However, this cannot be reached without accepting the travail of theoretical work. Still, the surplus value would be considerable, as hopefully has become apparent from the previous chapters.

References

1. Peschl MF (2001) Constructivism, cognition, and science-an investigation of its links and possible shortcomings. Found Sci 6(1–3):125–161
2. Krohn W, Küppers G (1990) Selbstreferenz und Planung. In: Niedersen U, Pohlmann L (eds) Selbstorganisation und Determination. Selbstorganisation. Jahrbuch für Komplexität in den Natur-, Sozial- und Geisteswissenschaften. Duncker & Humblot, Berlin, pp 109–128
3. Fränzle O (2001) Geleitwort. In: Ratter BMW (ed) Natur, Kultur und Komplexität. Adaptives Umweltmanagement am Niagara Escarpment in Ontario, Kanada. Springer, Berlin u.a., pp vii–xi
4. Egner H (2008a) Gesellschaft, Mensch, Umwelt - beobachtet. Ein Beitrag zur Theorie der Geographie. Erdkundliches Wissen. Franz Steiner, Stuttgart, p 208
5. Petley DN, Higuchi T, Petley DJ, Bulmer MH, Carey J (2005) Development of progressive landslide failure in cohesive materials. Geology 33(3):201–204
6. Bell R (2007) Lokale und regionale Gefahren- und Risikoanalyse gravitativer Massenbewegungen an der Schwäbischen Alb. PHD Thesis, University of Bonn, Bonn, p 270
7. Semenza E, Ghirotti M (2000) History of the 1963 Vaiont slide: the importance of geological factors. Bull Eng Geol Environ 59(2):87–97
8. Prigogine I, Stengers I (1990) Entwicklung und Irreversibilität. In: Niedersen U, Pohlmann L (eds) Selbstorganisation und Determination. Selbstorganisation. Jahrbuch für Komplexität in den Natur-, Sozial- und Geisteswissenschaften. Duncker & Humblot, Berlin, pp 3–18
9. Pohlmann L, Niedersen U (1990) Dynamisches Verzweigungsverhalten bei Wachstums- und Evolutionsprozessen. In: Niedersen U, Pohlmann L (eds) Selbstorganisation und Determination. Selbstorganisation. Jahrbuch für Komplexität in den Natur-, Sozial- und Geisteswissenschaften. Duncker & Humblot, Berlin, pp 63–82

10. Niedersen U, Pohlmann L (1990) Komplexität, Singularität und Determination. Die Koordination der Heterogenität. In: Niedersen U, Pohlmann L (eds) Selbstorganisation und Determination. Selbstorganisation. Jahrbuch für Komplexität in den Natur-, Sozial- und Geisteswissenschaften. Duncker & Humblot, Berlin, pp 25–54
11. Luhmann N (1989) Politische Steuerung: Ein Diskussionsbeitrag. In: Hartwich H-H (ed) Macht und Ohnmacht politischer Institutionen. Westdeutscher Verlag, Opladen, pp 12–16
12. Hertz H (1894) Die Prinzipien der Mechanik in neuem Zusammenhange dargestellt. Leipzig
13. Ratter BMW, Treiling T (2008) Komplexität—oder was bedeuten die Pfeile zwischen den Kästchen? In: Egner H, Ratter BMW, Dikau R (eds) Umwelt als System—System als Umwelt? Systemtheorien auf dem Prüfstand. oekom, München, pp 23–38
14. Bühl WL (1987) Grenzen der Autopoiesis. Kölner Zeitschrift für Soziologie und Sozialpsychologie 39:225–254
15. zur Lippe R (1994) Denken und Leben. Essay zur Einführung von Rudolf zur Lippe. In: Maturana HR (ed) Was ist Erkennen? Piper, München, pp 7–23
16. Maturana HR (1994) Was ist Erkennen? Piper, München, p 244
17. Egner H (2008) Planen, beeinflussen, verändern… Zur Steuerbarkeit autopoietischer Systeme. In: Egner H, Ratter BMW, Dikau R (eds) Umwelt als System—System als Umwelt? Systemtheorien auf dem Prüfstand. oekom, München, pp 137–154
18. FEMA (2004) Hurricane Pam Exercise Concludes. http://www.fema.gov/news/newsrelease.fema?id=13051. 19 Nov 2009)
19. Fragilecologies Archives (2006) The Perfect "Storm Scenario": The Hurricane Pam Exercise. http://fragilecologies.com/blog/?p=601. 19 Nov 2009)
20. Ratter, BMW (2001) Natur, Kultur und Komplexität. Adaptives Umweltmanagement am Niagara Escarpment in Ontario, Kanada. Springer, Berlin u.a., p 315
21. Prigogine I, Stengers I (1981) Dialog mit der Natur. Neue Wege naturwissenschaftlichen Denkens. Piper, München, Zürich, p 314
22. Schrödinger E (1944) What is life? The physical aspect of the living cell. Cambridge University Press, New York, p 32
23. Egner H, von Elverfeldt K (2009) A bridge over troubled waters? Systems theory and dialogue in geography. Area 41(3):319–328

Chapter 11
Meeting the Challenge ... An Approach to a Geomorphological System Theory

No-one goes further than (s)he who does not know where (s)he is going.

Goethe, translation by KvE

At the beginning I raised the question whether system theoretical thinking is a challenge for geomorphology. Unambiguously, this question has to be answered in the affirmative. This is not only due to the fact that *per se* a system theoretical foundation of a science is a challenge: After all, modern system theories force scientists to think in loops, as simple causalities are questioned per definition. The specific challenge for geomorphology, however, goes further than this. It is rooted in the epistemological imperative of empiricism that strongly (and nearly utterly) determines geomorphological research. This results in a marginalisation of geomorphological theoretical work and research. This dissertation thesis represents an explicit alternative draft to this research practice. It stresses that theoretical research offers a considerable surplus value. Just as the Austrian composer Anton Bruckner (1824–1896) said: (S)He who wants to build high towers has to give special attention to the foundation. Now, what about this foundation within geomorphology?

11.1 The Challenge

A main objective of this thesis is to make explicit the theoretical basis of geomorphology. Thus, the challenges for geomorphology can be clearly stated. The basic hypotheses for this work, as presented in chapter 1, refer to this foundation of geomorphology. The resulting questions can now be answered:

Firstly: How stringent and structuring are current basic definitions in geomorphology?

As I have demonstrated in Chap. 3, geomorphological definitions of 'system' are based on *common sense*. According to the conventional definitions, systems can be defined and delimited arbitrarily and according to the respective research focus.

K. von Elverfeldt, *System Theory in Geomorphology*,
Springer Theses, DOI: 10.1007/978-94-007-2822-6_11,

The environment of the system is simply 'everything else', i.e. what is not considered and viewed. The definitions and thus also geomorphological research are focussed on the openness of systems. This results in the study of input–output-relations of the system with its environment. Hence, geomorphological system research is a mere analysis of flows of matter and energy. This only has the potential of description, not of explaining the structuring (=creation of landforms) of the earth's surface. In this manifestation, geomorphological system theory is a mere terminological transformation of traditional landscape concepts.

Secondly: How is the world perceived within geomorphology and how big is the organizing potential of our background theory?

It is one of the main tasks of science to offer schemes for the description of the world. This then transforms the observed phenomena into a specific order. As the geomorphological system theory does not offer a coherent rule according to which systems can be delimited, and as the openness of systems is focussed, geomorphic systems appear to be embedded into their environment. Thus, the first law of geomorphology is as follows: Everything is connected to everything else. However, it is nearly impossible to attribute a scientific ordering function to this premise. The same is true for the current classification approach within geomorphology. It tries to distinguish system types by the complexity of their structures. At best, this is an instruction for reductionism, but only has a limited ordering potential.

Thirdly: How firmly is geomorphological theory and practice rooted in physics?

When studying physical—and thus also geomorphic—systems, three levels can be distinguished that cannot be reduced to each other: mechanics, thermodynamics, and non-linear thermodynamics. Mechanics is a study on the level of particles, and thus does not allow for statements on the system as a whole. Thermodynamics, in contrast, describes the whole system with statistical means. Research objects of classical thermodynamics are those linear systems that are close to or in thermodynamic equilibrium, that is, static systems. If statements on the evolution of a system are to be made, these have to be based on the physics of non-linear thermodynamics. As I have shown from Chap. 5 onwards, the geomorphological adaptation of this physical basis resembles shaky ground rather than a solid foundation. The different, incompatible levels of investigations—mechanics, thermodynamics, and non-linear thermodynamics—are blended. This results in statements that are not congruent with the physical basis. A striking example for this is the equilibrium thinking that is still present in geomorphology. In most cases, it assumes a mechanical equilibrium, and from this assumption the system as a whole is assessed. Consequently, stable structures are regarded as an expression of an equilibrium (defined in manifold and contradictory terms). However, according to the theory of dissipative structures these stable structures are a characteristic of self-organisation and thus of an inherent non-equilibrium.

This contradiction with physics is probably the gravest consequence of the hostility towards theoretical research in our discipline.

Fourthly: Can geomorphologists read the landscape?

The answer is this: not, if ‘reading’ is understood as the (‘right’) recognition of a reality. Observation always depends on theory. As it were, we need a concept according to which we can distinguish ‘the things’ from each other. In other words, we always have a classification concept in the back of our heads that is continuously activated. This then helps us to recognize, for example, a table, a chair, a river, a rock glacier, a rockslide as such (and to classify it as such). This is why we can distinguish all these ‘things’ from everything else. In the background, the theory is constantly retrieved, and thus it is the theory—and not the reality!–, which determines what we see. However, we do not only rely on certain, largely unconscious presumptions, but also on our recognitional apparatus that is simply subject to certain limitations. After all, in order to survive, it is sufficient that our species has a general idea of what ‘really’ ‘is’ around us. Anthropocentrically, it could even read: It is especially economical and reasonable to ‘biologically equip’ mankind with just as much as (s)he needs for survival. Thus, from the fact that we always perceive precisely what we are perceiving, it cannot be concluded that we necessarily perceive everything.

Fifthly: If the theoretical basis changes, what does this mean for empirics?

With the examples of system control, management, and prediction, I have demonstrated the possible implications of a geomorphological system theory, which is strongly anchored within the paradigm of self-organisation. Against the background of non-linear thermodynamics, especially the theory of dissipative structure of Ilya Prigogine, it is thus impossible to make long-term predictions about future system states. This is due to the self-organisation of dissipative systems. In certain phases, these systems thus evolve independently from external conditions. From this it becomes apparent that in these phases they cannot be controlled, or at best can be barely controlled. Furthermore, depending on the respective system state, one and the same attempt of control can have completely different effects. In principle, geomorphologists are already used to this effect, but this effect can now be explained. Take the example of a slope beyond instability: Possibly, drainage is useless—the movement will just continue. Another slope, albeit also in the process of sliding, may be stabilised by exactly the same measure. This exemplifies the planning dilemma—we are holding the wolf by the ears: Steering and control does not necessarily (to be more exact: only rarely does) result exactly in the intended consequences. However, not to do anything is no option either. Flexibility could be the loophole. That is, we need to turn away from the assumption of rigid and inflexible process structures, as this obviously leads to rigid and inflexible management structures, too. Adaptive management seems to be such a flexible approach, which focusses the self-regulation and -organisation of systems. Hence, it provides for the continuous adaptation of the measures to a system that continuously changes its behaviour.

It is rather obvious that interdisciplinary cooperation improves the chances of successful adaptive management. However, so far, most of the attempts of inner-geographical cooperation have failed or have not lead to the anticipated results. One reason for this has been worked out in Sect. 10.2: Due to the immensely differing initial theoretical assumptions of the respective part-disciplines an understanding is virtually impossible. Yet again, second order system theories offer an approach to a solution: They are already utilized both in natural and social sciences. This seems especially important, as a successful inter- and intra-disciplinary cooperation is needed in order to tackle the big problems of the twenty first century, e.g. global (environmental) change.

11.2 An Approach to a Geomorphological System Theory

I have introduced second order system theories as an alternative offer to the inconsistent theoretical foundation of geomorphology. These second order system theories encompass approaches from biology, physics, and sociology. However, second order system theories are an arduous and especially uncomfortable task (cf. [1, p. 30]), as this thinking forces one to give up nearly all former opinions and conceptions. But this is not the only reason why system theoretical abstraction means a loss of securities [2, p. 146]. The main reason probably is the realisation that all recognition is subjective, and that all recognition has blind spots. Thus, the claim of truth is no longer sustained. Consequently, the question is this: Why should we as geomorphologists be willing to impose these exertions and uncertainties upon ourselves?

Based on the observation of the discipline of geomorphology (at least) eight arguments can be summarized, illustrating why an advancement of geomorphological system theory appears to be highly attractive:

1. A clear separation of organisation and structure offers an explanation why radical changes as well as continuous changes can take place whilst the organisation and the *unity* of the system stay the same (also cf. [3]). The scope of possible geomorphological analyses could be largely expanded. Furthermore, these thoughts are consistent with the theory of thermodynamics and with current concepts of self-organisation and non-linearity. This, however, does not apply to the present system theoretical basis of geomorphology.
2. The strict separation of organisation and structure offers an understanding why quite a few geomorphic systems (such as glaciers and rivers), under changed conditions, just ‘go on as before’. They only react to a lesser extent to the changes—the structure of the system simply does not provide for this. Even a catastrophic (from human perspective) change in the sense of a disappearance of a system (e.g. of a glacier), or in the sense of a radical change of system structure (e.g. avulsions, landslides) basically mirrors that the system only has the choice between being adapted or simply being not adapted. Thus, environmental conditions lose their explanatory power for system behaviour (also cf. [4]).

3. The renunciation of the reductionist approach on systems in equilibrium (cf. Sect. 8.2) opens the view for the recognition of spatial structures (and also: temporal structures) (also cf. [5]).
4. If we accept self-reference as constituting characteristic of systems, this results in a shift of geomorphological methods. After all, if system operations/processes (operative closeness) are focussed, the specification of input–output-relations becomes obsolete: It is not necessary anymore to analyse the dependency of the system on 'everything else'. Thus, the complexity of our world is reduced considerably, thereby fulfilling a claim to science (cf. [6]). Additionally, the assumption of a teleology of nature can be omitted, which—in the face of the paradigm of self-organisation—seems to be rather old-fashioned anyway. An autopoietic system does not have any purpose, it only 'keeps on operating' and thereby continuously reproduces itself.
5. If second order system theories are applied to geomorphology, mechanistic equilibrium thinking may eventually lose its fascination for geomorphologists. Furthermore, second order system theories agree with (non-linear) thermodynamics. Non-linear systems can be long-lasting stable structures, and at the same time they are capable of substantial change at crossroads (i.e. bifurcation points). Then, they may exhibit a new behaviour—but, at the same time, this does not imply a change in its fundamental way of operation. A good (everyday) example by Maturana has already been mentioned before: After a structural change a chair might be wobbly, or—in order to use a geomorphological example—a river might leave its bed (avulsion), and both changes might have rather 'catastrophic' dimensions. However, the mode of operation (the processes) would still be the same in both cases: A chair thus would still be (recognisable as) a chair, and the river would be a river.
6. Self-organisation, complexity, self-organised criticality, to name only a few, are already central concepts of geomorphology. These could be easily integrated, and no 'theoretical crutch' would have to be employed. This, then, would also represent a farewell to the mere terminological transformation of 'old commonplaces' (cf. [7, p. 94]).
7. The discussion of the concept alone would initiate an overdue debate on discipline-specific theories, as well as on the philosophy of science, and on epistemology. This can only be for the benefit of the discipline. After all, the concepts of equilibrium and of 'everything is connected to everything else' are a characteristic of pre-scientific thinking rather than a characteristic of scientific spirit [7, p. 95]. Scientific approaches are defined by deliberate selection and isolation of interrelations. Geomorphology finally has to rise to this challenge.
8. By integrating core-elements of second order system theories into geomorphology, a scientific connectivity to research fields of human geography (as well as other social sciences) would exist on the level of theories. In the face of the so-called ecological crisis and of apparent global change it becomes increasingly obvious that we need integrative concepts if we want to tackle such big questions. The troubled waters between natural and social sciences

have to be bridged (e.g. cf. [4]). With a sophisticated system theoretical foundation, geomorphology could fruitfully contribute to this task.

Thus, geomorphology faces the challenge of a paradigmatic change: A change towards the paradigm of self-organisation and thus towards second order system theories. But what are the possible consequences of such a paradigmatic change for geomorphology? With Luhmann [8, p. 23] and Varela (in [9, p. 143]) three theses can be named:

1. Theory undergoes a conversion away from focussing the entity of a whole (e.g. a glacier system) within a bigger entity (e.g. a mountain range or the world as a whole) towards focussing the difference of a specific system and its environment. In this way of thinking, system and environment cannot be thought of without the respective other. Thus, the boundaries between systems and their respective environments are focussed. This also means that one must ask how the system organises itself and how it is able to have relations with its environment.
2. Thus, interactions are no longer an element of a definition of a system, especially as they are not specific to any system. Interactions are nothing other than perturbations, which may (or may not) stimulate the system to evolve according to its own dynamics and structure.
3. A completely new understanding is needed of what can be considered as system elements: These are no longer single compartments, variables, or sub-systems. Instead, it is the self-referential operations, which are the system elements. It is the self-referential operations that initially bring forth the system (keyword autopoiesis), or which delimitate it (keyword self-reference). In other words: The processes as constituting (and thus defining) system elements are at the centre of attention.

This is still a new and unfamiliar approach for geomorphology. It describes systems as a coherent interplay of processes, which create structures that may be temporarily stable. At the same time, these temporarily stable structures have nothing to do with equilibria of any kind (also cf. [10]). For example, an oversteepened slope that starts to move because of this oversteepening, changes its structure. But this does not imply that the slope gets out of equilibrium or that it strives for a new equilibrium by adapting to environmental conditions. In the perspective of second order system theories, a landslide can be seen as an expression of the plastic system structure. At the same time, the existence of this (or another) structure expresses that the system possesses free energy. Thus, it is far from thermodynamic equilibrium. If these structures are stable—if, for example on a slope landslides are not occurring—this means nothing more than exactly this: The structures are stable. It thus in no way means that the system is in equilibrium of any kind.

Of course, environmental conditions may also influence a system in steady state, i.e. when it exhibits stable structures. But they can influence a system only as long as its internal structures and its self-organisation permit this. Thus, it can be explained why, on the one hand, we perceive systems that are influenced by their environment or that structurally adapt to environmental conditions. And it can be explained why we, on the other hand, know of systems that do not change structurally over long timescales, despite striking environmental changes. However, the reverse is also true: Systems, which change their structures without any apparent change in environmental conditions. For example, what has been termed as the 'long' reaction time of rock glaciers can be explained in a completely different way: The self-organised and structurally-determined system maintains itself in a specific state. Environmental conditions only become important when the 'order through fluctuations' collapses [e.g. because specific gradients (of temperature etc.) are becoming too weak] and the system moves into a different stage or ceases its existence.

The new perspective of second order system theories thus can be summarized as being process-oriented. The development of order, that is, of structures, is focussed. The current geomorphological system theory, in contrast, concentrates on interactions and system components in given structures. The difference and the consequences are striking: Second order system theories offer the potential to study structure-building processes and to elicit the relations between form and process. Hence, what more can a geomorphologist wish for from a theory?!

References

1. von Glasersfeld E (2002) Ernst von Glasersfeld im Interview mit Reinhardt Voss. "... es ist eine anstrengende und vor allen Dingen ungemütliche Sache". In: Reinhardt V (ed) Unterricht aus konstruktivistischer Sicht. Die Welten in den Köpfen der Kinder. Luchterhand, Neuwied, pp 26–32
2. Egner H (2008) Gesellschaft, Mensch Umwelt—beobachtet. Ein Beitrag zur Theorie der Geographie. Erdkundliches Wissen. Franz Steiner, Stuttgart, p 208
3. Mingers J (2002) Can social systems be autopoietic? Assessing Luhmann's social theory. Sociol Rev 50(2):278–299
4. Egner H, von Elverfeldt K (2009) A bridge over troubled waters? Systems theory and dialogue in geography. Area 41(3):319–328
5. Hess B, Markus M (1986) Chemische Uhren. In: Dress A, Hendrichs H, Küppers G (eds) Selbstorganisation. Die Entstehung von Ordnung in Natur und Gesellschaft. Piper, München, Zürich, pp 61–80
6. Egner H (2010) Theoretische Geographie. Wissenschaftliche Buchgesellschaft, Darmstadt, pp 144
7. Hard G (1973) Zur Methodologie und Zukunft der Physischen Geographien an Hochschule und Schule. Möglichkeiten physisch-geographischer Forschungsperspektiven. Geog Z (61):5–35 (hier aus: Hard G (2003) Dimensionen geographischen Denkens. Aufsätze zur Theorie der Geograpie, vol 2. Osnabrück, pp 87–111)
8. Luhmann N (1986) Systeme verstehen Systeme. In: Luhmann N, Schorr KE (ed) Zwischen Intransparenz und Verstehen. Frankfurt a.M, pp 72–117
9. Simon FB (1997) Kreuzverhör: Fragen an Heinz von Foerster, Niklas Luhmann und Francisco Varela. In: Simon FB (ed) Lebende Systeme: Wirklichkeitskonstruktionen in der systemischen Therapie. Suhrkamp, Frankfurt/Main, pp 131–147
10. Jantsch E (1987) Erkenntnistheoretische Aspekte der Selbstorganisation natürlicher Systeme. In: Schmitt JS (ed) Der Diskurs des Radikalen Konstruktivismus. Suhrkamp, Frankfurt/Main, pp 159–191

Chapter 12
Summary

Geomorphology understands itself as applied science, within which—so far—the theoretical foundation only plays a minor role. Against this background, this thesis wants to contribute to a system theoretical foundation of geomorphology. It aims at uncovering the surplus values of theoretical research for the discipline. Methodically, it is based on the observation theory of Heinz von Foerster. Foerster's understanding of observation differs fundamentally from the everyday (scientific) observation: he defines it as an act of distinction. The criteria upon which an observation, and thus the resulting knowledge, is based, thus can only be recognised in a second step. This second step is called second order observation, and it is this second step, which is taken in this thesis.

It is shown that, within geomorphology, the theoretical foundations are rarely questioned, which gives rise to five problem fields:

(1). Absent coherence of the basic assumptions and concepts. Thus, the research object remains diffuse.
(2). Lack of analytical structure. As everything is seen as being connected with everything else, system research focuses on input–output-relations rather than on the inner dynamics and structure of systems: The system itself remains a black box.
(3). The geomorphological system theory is widely incompatible with physical basics, as different levels of observation are blended. As a result, scientific statements often inadvertently exceed the scope of research results.
(4). Equilibrium thinking has already moulded geomorphological research for decades despite the fact that it conflicts with thermodynamics.
(5). Despite a relatively new focus on complexity and non-linearity, equilibrium thinking has continued until recently, though without possessing any theoretical foundation.

K. von Elverfeldt, *System Theory in Geomorphology*,
Springer Theses, DOI: 10.1007/978-94-007-2822-6_12,

The reception of current approaches from biology (e.g. self-reference and autopoiesis) as well as thermodynamics (non-equilibrium thermodynamics) is suggested for resolving these difficulties and for achieving a coherent background theory. In a second step, the resulting consequences for geomorphology are discussed. The thesis concludes by drawing the outlines of a consistent, physically (thermodynamically) sound geomorphological system theory, which shows a high scientific connectivity with our neighbouring disciplines.

Index

K. von Elverfeldt, *System Theory in Geomorphology*,
Springer Theses, DOI: 10.1007/978-94-007-2822-6,